ON THE SHOULDERS OF GIANTS

THE HISTORY OF SCIENCE IN THE EIGHTEENTH CENTURY

Ray Spangenburg and Diane K. Moser

On the cover: Lazzaro Spallanzani performs an experiment on digestion in birds. (Figuier, *Vie des Savants*, 1870 [Courtesy, North Dakota State University Library])

The History of Science in the Eighteenth Century

 For information contact:
Facts On File, Inc.
460 Park Avenue South
New York NY 10016
USA

Library of Congress Cataloging-in-Publication Data

Spangenburg, Ray, 1939–
The history of science in the eighteenth century / Ray Spangenburg and Diane K. Moser.
p. cm. — (On the shoulders of giants)
Includes bibliographical references and index.
ISBN 0-8160-2740-4
1. Science—History—18th century. I. Moser, Diane, 1944– . II. Title. III. Series: Spangenburg, Ray, 1939– On the shoulders of giants.
Q125.S736 1993
509′.033—dc20 92-41500

A British CIP catalogue record for this book is available from the British Library.

Facts On File books are available at special discounts when purchased in bulk quantities for businesses, associations, institutions or sales promotions. Please contact our Special Sales Department in New York at 212/683-2244 (dial 800/322-8755 except in NY).

Text design by Ron Monteleone
Cover design by Semadar Megged
Composition by Facts On File, Inc./Robert Yaffe
Manufactured by the Maple-Vail Book Manufacturing Group
Printed in the United States of America

10 9 8 7 6 5 4 3 2 1
This book is printed on acid-free paper.

Dedicated to the memory of
Voltaire
(François Marie Arouet)
(1694–1778)
and the spirit of the Enlightenment

CONTENTS

ACKNOWLEDGMENTS

So many people have been generous with their time, talents and expertise in helping us with this book—both as we wrote it and in the past when it was only a dream. We'd like to thank them all—and especially: Gregg Proctor and the rest of the staff at the branches of the Sacramento Public Library for their tireless help in locating research materials. Dorothea Nelhybel at the Burndy Library for her willingness to help with photos and illustrations despite the fact that her library was in the middle of moving. Beth Etgen, educational director at the Sacramento Science Center, and her staff, as well as science historian Karl Hall, for kindly reading the manuscript and making many helpful suggestions. And for their help with illustrations: Andrew Fraknoi of the Astronomical Society of the Pacific, Leslie Overstreet of the Smithsonian Institution Libraries, Diane Vogt-O'Connor of the Smithsonian Archives, Jan Lazarus of the National Library of Medicine, Clark Evans of the Library of Congress, R. W. Errickson of Parke-Davis and Cynthia M. Serve of Bausch & Lomb. Thanks also to Facts On File's editorial team, especially Nicole Bowen for her intelligent, energetic and up-beat management of the project; Janet S. McDonald for her eagle-eyed passion for detail; and James Warren for his vision and encouragement in getting it started. And to many others, including Jeanne Sheldon-Parsons, Laurie Wise, Chris McKay of NASA Ames, Robert Sheaffer and Bob Steiner, for many long conversations about science, its history and its purpose.

P R O L O G U E

THE STORY SO FAR: THE COPERNICAN SYSTEM AND THE SCIENTIFIC REVOLUTION

People have always looked for knowledge in various ways, delving for explanations and searching their world for answers. And people have always used the richness of their minds—their imaginations and their faith—to bridge what seemed impossible to understand. From the beginning of time they have told stories of gods and superhuman men and women who forged the forces that made the world work, the Sun rise, the rains come and the harvests grow. And shamans and priests fashioned rituals by which they sought to control pestilence and disease and to drive away drought.

However, from earliest times human beings have also used their powers of observation and calculation. Very early, people in several areas of the world—including China and the Mesopotamian basin—began to develop the tools of mathematics, language and writing. And they developed more quantitative and objective approaches to explaining. By 1800 B.C., the Sumerians in the Middle East and the Babylonians, who succeeded them, had made many accurate observations about the stars and planets and had developed numerical systems and ways to keep records. Cuneiform and pictograph writing systems began even earlier and by 1300 B.C., the Phoenicians had alphabets based on earlier systems developed by the Egyptians and the Babylonians. And by the time of the ancient Greeks, in the last four or five centuries B.C., analytical methods, logic and geometry began to be finely honed. Several Greek thinkers, especially, began to explore more nonmystical explanations. And Aristotle (384–322 B.C.), perhaps the greatest Greek thinker, and several of his followers, including a man named Ptolemy, also devised the first integrated natural theory about how the universe worked, with intricate visions of nested spheres that held the planets and the stars and circled around the Earth.

Greek ideas about mathematics, philosophy and science reigned supreme up to the fall of the Roman Empire, when many manuscripts were burned as cities were sacked by barbarian invasions. But some were preserved, largely by Islamic priests, and later passed into hands of monks in the monasteries of Europe. Partly as a result of this preservation, and partly because no one questioned it, Aristotle and Ptolemy's concept of a universe composed of concentric spheres revolving around the Earth remained the best explanation of the cosmos anyone came up with for some 14 centuries.

Then, in a daring sweep of imagination, the Polish theorist Nicolaus Copernicus revised that theory in the 16th century by placing the Sun, not the Earth, at the center of the circling planets. Called the Copernican system, this Sun-centered view of the Solar System (and actually, the universe, at the time) turned every doctrine and assumption about the world and humanity's place in it upside-down. When two more great astronomers, the Danish observer Tycho Brahe (1546–1601) and his one-time assistant from Germany, Johannes Kepler (1571–1630), added their own observations and calculations to verify Copernicus's ideas, most people thought they, too, were heretics. And when the Italian astronomer Galileo Galilei (1564–1642) defended the idea in his book *Dialogue on the Two Chief Systems of the World*, published in 1632, he was tried by the Catholic church and his book was banned—put on the Index, where it stayed until 1835. (Not until 1992, in fact, did the pope declare that Galileo was not in error.) Even though careful inquiry and reasoning supported Copernicus's idea, a Sun-centered solar system didn't make sense if one took certain passages of the Bible literally. Theologians and their followers pointed significantly to the Old Testament story in which God stopped the movement of the Sun across the sky so that the people led by Joshua could win a crucial battle. The Sun, then, must be moving around the Earth and not vice versa. To resolve the conflict between nature observed and the Bible's authority, church authorities recommended keeping faith with the Bible. Church doctrine also maintained that humanity was created in God's image and that therefore God would not have created the universe in a way that did not revolve around the Earth. And so a battle raged between those who wanted to rely on human ability to find things out through the use of the mind and the senses and those who preferred to depend on traditional authority for answers.

Nearly everyone today takes Copernicus's idea for granted that the Earth revolves around the Sun. But the stir his ideas caused in his time, and Galileo's, is a reminder of how hard it is to let go of old ways of thinking in the light of new facts. What made arguments for the Copernican system so powerful at the time, even against the spiritual and political strength of religion and tradition, was the birth of a new methodology, spawned in large part by the experimental work of Galileo, an extraordinarily effective approach to problem solving that came to be known as the scientific method (explained in more detail in the Appendix). Scientific method calls for explanations based on observation, on collected facts and measurements—not on reasoning alone or on emotional reactions, visions, hearsay or faith. And the scientific method accepts only explanations that can be verified over and over by experiments. The results of these experiments, summarized in theories, can then be used to make predictions about other phenomena not yet observed. Then when the opportunity arises to test one of these predictions, the results of that test can be compared to the predicted results. In this way, the results of experiments and observations are constantly used to modify existing theories—which is what makes science a "self-correcting" process. This fundamental change in the way people looked at the world, which took place in the period just preceding the first years covered in this book, is commonly referred to as the scientific revolution.

Many of the most important scientific discoveries in the 17th century had centered in Italy. Even the Polish astronomer Copernicus had trained there in the 1490s, and while Tycho and Kepler had come from the more northern regions of Denmark and Germany, the great names in the life sciences either trained in Italy or were born there. Among many others, Andreas Vesalius (1514–64), the Flemish father of human anatomy, William Harvey (1578–1657), the English physician who made breakthrough progress in the area of blood circulation, and Marcello Malpighi (1628–94), the Italian physiologist who discovered capillaries in frogs' lungs, studied in Italy.

But by the end of the century the focus of growth had begun to shift north to France and England. In England, especially, economic dependence on far-flung colonies in North America, Africa and Asia created an urgent need for reliable knowledge about navigation. The English government—understanding that knowledge of the stars and their positions was the key to navigation—created the position of Royal Astronomer and John Flamsteed (1646–1719) was appointed to the post. Edmund Halley (1656–1742), Christopher Wren (1632–1723) and Robert Hooke (1635–1703), among others, formed a society of scientists called the Royal Society in 1662. (Its motto, Nullius in Verba—take nobody's word; see for yourself—set the stage for the century to come.) The greatest of all English scientists, Isaac Newton, became a member of the Society in 1671, and in 1687 he published a book called *Principia mathematica* that synthesized the ideas of the scientific revolution and set forth basic principles of the universe, "laws" of motion that nature followed, in a more cohesive way than anyone ever had before.

In what are now known as Newton's three laws of motion, the great English synthesizer mathematically demonstrated principles that Galileo explored with some of his experiments:

1. the principle of inertia (a body at rest remains at rest and a body in motion remains in motion at a constant velocity as long as outside forces are not involved);
2. definition of a force in terms of mass and acceleration; and
3. the famous statement that for every action there is an equal and opposite reaction (the principle that governs the behavior of rockets).

By the beginning of the 18th century, the scientific revolution, based on the work of such intellectual giants as Galileo and Newton, had established a new way of perceiving nature and had forever altered humanity's understanding of itself and the universe.

The History of Science in the Eighteenth Century is one of a series of five books, called On the Shoulders of Giants, which looks at how people have developed the methods of science as a system for finding out how the world works. We will look at the theories they put forth, sometimes right and sometimes wrong. And we will look at how we have learned to test, accept and build upon those theories—or to correct, expand or simplify them.

We'll also see how scientists have learned from others' mistakes, sometimes having to discard theories that once seemed logical but later proved to be incorrect, misleading or unfruitful. In all these ways these men and women—and the rest of us as well—have built upon the shoulders of the men and women of science, the giants, who went before them.

THE EIGHTEENTH CENTURY—A TIME OF REASON AND REVOLUTION

Variously called the Enlightenment or the Age of Reason, and even, sometimes, the Age of Confidence, the 18th century was a time of enormous optimism about the power of human reason to discover all ultimate truths and overcome all problems, intellectual, philosophical and social. From the great thinkers of the scientific revolution—Nicolaus Copernicus, Galileo Galilei and Isaac Newton—they had inherited a key that seemed capable of unlocking all the secrets of Nature. And at the dawn of

Newton in his laboratory studying the nature of light (Louis Figuier: *Vies des savants illustres depuis l'antiquité jusqu'au dix-neuvième siècle*, vol. V: *Savants du XVIIIe siècle*, 1870)

the 18th century the world—in fact, the entire universe—began to seem altogether knowable.

Of course, the farmer tilling his fields, the midwife calmly performing her duties and the shopkeeper selling his wares had little time left in their crowded days to contemplate the nature of the universe or humankind's place in it. But even the farmer, the midwife and the merchant, along with millions of others, found that the spirit of the times was changing. Not everyone liked the transformation, and not everyone understood it or accepted it, but by the end of the century it had touched the lives of virtually everyone in the Western world. The 18th century's three great revolutions—the American, the French and the Industrial—were all touched off by this new outlook, and its impact helped to define the world we live in today.

At the heart of this change stood Newton, his *Principia mathematica* and the natural laws of gravity and motion he uncovered. Nature, Newton had demonstrated, played by rules. And the rules appeared to be regular and identifiable. Nature was no longer seen as a capricious and unpredictable force; instead it was governed by basic laws that it always obeyed. For the thinkers of the 18th century Newton's discovery of the existence of these laws disclosed a clockwork universe. They saw the world as a machine. And they thought that one could understand the world-machine in the same way that a skilled craftsperson who had never seen the inside of a clock might take one apart, examine its parts and their relationships, apply common sense and reason and comprehend how it worked.

Newton, building on the work of Copernicus, Tycho Brahe, Kepler and Galileo before him, gave final proof to the power of the scientific method. And this new approach to nature had caused an avalanche of consequences in his time. Not only did 17th-century scientists set straight many ancient distortions about the way the physical world worked, but some—most notably the English physiologist William Harvey—also used the new method to discover the internal mechanisms of the human body. And, while progress in biology occasionally stumbled when scientists applied too strict a "Newtonian" and "mechanistic" interpretation to human physiology, 18th-century thinkers saw that it was possible to use the scientific method of observation and experimentation to answer many difficult questions successfully. No longer would answers have to come from the books and writings of ancient thinkers or the unchallengeable word of religious authorities.

This new outlook liberated the thinkers of the 18th century from the authority of ancient Greek theories that had never been tested by experiment or observation. Knowledge of the world, they realized, could be acquired, not passively from Greek and Latin manuscripts and wizened scholars, but from the active, curious and disciplined engagement of their minds applying rational thinking, combined with experimentation and observation, to a variety of problems.

Was there room in this new "enlightened" view for religion? Newton himself was troubled by this question. Devoutly religious, Newton never gave up his belief in God. He believed in fact that it was occasionally necessary for God to intervene to keep the Solar System working in its beautiful machinelike precision. Finally he concluded that the wonderful machinery of the universe, in all its clockwork precision, demonstrated the magnificent power and majesty of its creator. In the 18th century, after it was discovered by mathematical analysis that God's intervention was not necessary to keep the Solar System stable, many adopted a religious view called deism. This view held that God had created the world and all of its natural laws but, having finished the job, had left the machinery to run by itself. As Newton had foreseen, though, this view left some people wondering whether a clockwork universe required the existence of God at all. Consequently—and this possibility had troubled Newton—some of the enlightenment thinkers became atheists, denying the need or existence of any god. One group especially, the philosophes in France, strove to build a moral philosophy based not on revealed religion, but on human ethical thought. Their ranks included such great thinkers as Voltaire, Diderot and Montesquieu. So, while by far the vast majority remained theists, maintaining their allegiance to God and the traditional teachings of religion about miracles, Newton's legacy also introduced new doubts about fundamental issues of religion and philosophy.

It presented new questions about social structure and values as well. Scientists had discovered that nature had "natural laws." Was it possible that similar natural "laws" might govern all moral, social and political activities? If that was the case, many believed, then by applying rational thinking, those laws could also be discovered and turned toward the benefit of humanity. Philosophers such as Immanuel Kant (who coined the term *enlightenment* to describe the new dawn of reason and the intellectual brilliance of the time), David Hume, Gottfried Wilhelm Leibniz, François Marie Arouet (known as Voltaire), Jean-Jacques Rousseau and others sought to find a place for humanity in the new perspective brought by science to the world. Some, like Voltaire, who attended Newton's funeral, became vocal and dynamic champions of the new scientific-sociopolitical views. Writers such as the French journalist Denis Diderot, who supervised the first 28 volumes of the great *Encyclopédie*, brought the new ideas of science and philosophy to the general public.

In Paris, writers, philosophers and artists gathered at *salons* in the homes of wealthy and intelligent French women, who invited them to meet there and exchange views. It was a time when lively, impassioned and closely reasoned discussion flew among the philosophes, not only at the salons of

Paris, but also in lengthy correspondence in which a number of women also took part. Few women were permitted the education needed to contribute in more direct ways, but a notable exception in these circles was Emilie du Châtelet, whose translation of the work of Newton from Latin to French was responsible for its wide dissemination and much of its influence in France. Several women took an active part in the advances of science, as will be pointed out in the chapters of this book.

The new thinking of the Enlightenment burst like fireworks across the social, geographical and political boundaries of the Western world. Across the Atlantic in the American colonies it kindled the thinking of political leaders and molders of government such as Thomas Jefferson, Thomas Paine and scientist-statesmen-author Benjamin Franklin. Jefferson, as well as many other intelligent and well-read people of the 18th century, particularly felt the influence of the English philosopher John Locke (1632–1704). In his work *An Essay Concerning Human Understanding*, published in 1690, Locke declared that there were natural laws constructed by God to ensure the happiness and welfare of humanity. Paramount among those laws were the rights of life, liberty and property. Furthermore, Locke argued, while people originally lived in a state of nature free from restriction or restraint, the strong had violated that peaceful existence and taken unfair advantage over the weak. To defend themselves from such cruelty and enjoy their God-given rights people had chosen rulers under special agreements. In exchange for the ruler's protection and aid in holding onto their natural rights, they had agreed to abide by the ruler's decisions and dictates. If a ruler began to violate those natural rights, though, argued Locke, then the people had the right to disobey and, if necessary, to overthrow that ruler.

In England, the idea of natural human rights had already grown strong during the 17th century, as Parliament struggled to wrest power away from an autocratic monarchy. In 1628 Parliament forced King Charles I to sign the Petition of Right, which required approval of taxes by Parliament, established limitations on military law and held that no person could be imprisoned without a specific charge. In a civil war beginning in 1642, following several despotic moves by Charles, Parliament overthrew the king, replacing the monarchy with a republic headed by a Puritan general named Oliver Cromwell. But the revolution faltered at Cromwell's death in 1658, and in 1660 Parliament invited Charles II, son of the executed king, to take the throne. But Parliament took control again in 1688. In a move known as the Glorious Revolution, they offered the crown to the king's Protestant daughter Mary and her husband, William of Orange, over her Catholic brother James, who was next in line. And in 1689, the year before Locke published his *Essay*, Parliament drew up what is known as the English Bill of Rights. These events in England, combined with the influence of the scientific revolution, had great effect on government in the Western world.

In Britain they established parliamentary government, rule by law, limited monarchy and the protection of individual liberties.

Many French thinkers in the 18th century admired the English government and strove to develop philosophies of human government, education and society that reflected reasoned and just policies, based on an understanding of the essence of what it means to be human. These philosophes also left a powerful mark on the intellectual climate of the time. They promoted reform in France, reacting against the excesses of the absolute monarchy that ruled there. They objected to preferences given to nobility, who paid few taxes and received the best positions by right of birth. And the philosophes rankled against the special rights accorded the Catholic church in France, which paid no taxes and had the right to censor books, restricting their publication (as it did Galileo's *Dialogue on the Two Chief Systems of the World*).

Some, like the baron de Montesquieu, although himself a privileged aristocrat, focused on formulating governmental policies that would ensure individual rights for all. He suggested the principle of separation of powers, dividing authority among three branches of government, a system that greatly influenced the authors of the U.S. Constitution. He maintained that, by separating the legislative, executive and judicial powers, no one entity would ever have absolute authority.

The spirit of the time was so rich with ideas and new achievements that many of the philosophes, spearheaded by Denis Diderot and Jean d'Alembert, set out to collect and publish exhaustive information on science, technology and other fields of human knowledge. They called their work the *Encyclopédie*, and many of their contemporaries contributed articles, including Voltaire, Montesquieu and others. The first volumes appeared in 1751, but both the French government and church censors immediately banned it for encouraging a spirit of revolt, as well as promoting "moral corruption, . . . irreligion, and . . . unbelief." Diderot finally convinced the censors to release some of the volumes, but many of the articles had to be carefully worded to avoid arousing their anger again. Publication continued until 1772, and more than two dozen volumes finally appeared.

The philosophes supported Locke's view that people were born neither good nor evil, and that a child's mind was like a blank tablet on which no one has written. Nine out of 10 people, wrote Locke, "are good or evil, useful or not, [as a result of] their education." The idea had profound political implications—that people should have no special treatment by right of birth, only on the basis of their ability to contribute, which depended on their education. Society could produce good citizens through good education.

Across the ocean, Jefferson and his colleagues would borrow many of Locke's ideas when the American colonies sought to free themselves from the

Frederick II of Prussia considered himself a great friend of science. Here he welcomes the philosophe and encyclopedist Jean d'Alembert to Berlin. (Figuier: *Vies des savants,* 1870)

economic and political encumbrance of British rule. The American Declaration of Independence, which echoed many of Locke's ideas, the American Revolution of 1776, the American Bill of Rights and the democratic government that replaced the colonial regime all reflect Locke's influence.

In Europe, many 18th-century monarchs such as Prussia's Frederick II (1744-97) saw themselves as "enlightened rulers." Often though, while intellectually paying lip service to the new ideas on human rights sweeping around them, these rulers still held a tight grip on the lives of the people they saw as their subjects, showing attitudes held over from the medieval days of serfdom, when kings felt they owned their subjects and all they produced.

In France, the home of some of the greatest Enlightenment thinkers and scientists, change came later, and when it came it was violent, confused and bloody. Change was long overdue, and the French were not content to watch the English and the Americans enjoy the recognition of human rights while their society still labored under an outdated feudal system, plagued with economic inequality, special class privileges and the absolute rule of kings. At the same time that the philosophes were writing about equal rights, the French bourgeoisie, or middle class, was growing stronger, more well-to-do and more ambitious; the peasants and city workers lived in poverty; and the nobles sought even more power. France, meanwhile, faced a grave economic crisis. King Louis XIV, who died in 1715, had spent lavishly on luxurious living and wars. In his wake he left enormous debts. Those who followed, Louis XV and Louis XVI, borrowed more and spent even more lavishly. Because the church and the nobility were largely protected from taxation, and even most of the wealthy bourgeoisie were exempt, the only sources of revenue were the poverty-stricken peasants and city workers. In June 1789, the bourgeoisie tried to change the order of government by establishing a new legislative body called the National Assembly. But revolt was inevitable. On July 14, 1789, Parisians massed outside the Bastille, the great fortress prison that symbolized the monarchy's injustices and oppression. The angry mob stormed the prison, killing the commander and some of the guards; the Revolution had begun and the bloodshed did not stop for years. Louis XVI was sent to the guillotine in January 1793. His wife, Marie Antoinette, was executed the same year. A period known as the Reign of Terror began and between September 1793 and July 1794 at least 20,000 people (possibly even twice that many) were executed. But when the bloodshed ended at last and a new constitution was adopted in 1795, in France, too, the new constitutional government established many of the Enlightenment principles of the philosophes—including guarantees of individual rights. When Napoléon Bonaparte came to power in 1799 as first consul, he established a new system of laws, the Code Napoléon, based on the ideals of the Revolution.

Meanwhile, another kind of transformation—the mechanization of production known as the Industrial Revolution—began to change the shape of the economic world. The Industrial Revolution began about 1750–60, primarily in Great Britain, where a combination of factors, including advances in agriculture, converged to make dramatic changes possible—changes that had both positive and negative effects. In 1701 a man named Jethro Tull invented a mechanical seeder. A horse-drawn hoe soon followed, and, as a result, food production increased so dramatically that Great Britain could now feed its growing population without importing as much food as before, thus freeing up financial resources to import raw materials for other kinds of production. In its colonies, Britain had good sources of supply, especially for cotton, and good markets for manufactured products.

The textile industry, which relied at the time on hundreds of weavers and spinners isolated in their cottages, was the first to be dramatically

JOHN KAY AND THE FLYING SHUTTLE

One of the most influential of the earliest inventions of the Industrial Revolution was John Kay's flying shuttle. Patented in 1733, the machine was an improvement on the weaver's loom. Before Kay's invention, two weavers were needed to weave a wide piece of cloth. Each would stand on opposite sides of the loom and toss the shuttle back and forth across the width of the cloth. Kay's invention was simple. He mounted his shuttle on small rollers so that it would roll back and forth on a wooden rail. One weaver then could quickly move it from side to side by pulling on a chord that caused wooden hammers to hit it back and forth.

While the invention increased the speed of the weaving and made it possible to weave wider cloth than had been woven before, it also eliminated one of the two weavers previously needed for the job. That was not a happy situation for the weavers of Colchester where Kay first introduced his invention. They argued that Kay was trying to take their jobs away from them. Kay argued that more weavers would be needed to produce even more cloth than had been produced before, but that did not settle the problem of the weavers, who were afraid that until the demand for cloth increased, they would still be out of work.

Kay took his machine to Leeds, hoping to drum up some interest there, but his luck took another bad turn. The manufacturers at Leeds liked the machine but refused to pay him royalties for its use. His lawsuits against them proved practically useless when they united and fought him in the courts, costing him most of his money in heavy court expenses.

Disheartened and nearly broke, Kay returned to his native town of Bury, England in 1745. The flying shuttle had caught on and begun to replace weavers across England. Even though he was not making money with his invention, groups of angry laborers rioted and broke into his house, looting it and forcing him to flee. He escaped to Manchester, but eventually he was also forced to flee that city, hiding in a sack of wool. Feeling unsafe in England, Kay emigrated to France, where he died penniless.

transformed by a series of inventions. In 1733 John Kay invented the flying shuttle to speed weaving—with the result that one weaver could use all the thread produced by several spinners. In the 1760s James Hargreaves invented what he called a spinning jenny, a device that enabled a spinner to turn a series of spindles by using a single spinning wheel. The power that ran both these "machines" was still human power, but by 1769, Richard Arkwright had figured out how to use water power to run a spinning machine, and by 1785, Edmund Cartwright had invented a loom run by water power. Now the weak point in production was obtaining enough raw materials to supply the speeding spinners and weavers, especially cotton fiber, which holds its seeds tangled tightly among its strands. But in 1793, an American inventor, Eli Whitney, solved that problem with the invention of his cotton gin, which stripped the seeds out automatically.

By now the new machines began to be too expensive for individual weavers to rent or own, and with the new use of water power, it became necessary to have a source of running water nearby for weaving and spinning. And so, entrepreneurs began to build and organize factories for efficient production, purchasing the equipment, hiring workers and locating new markets for the goods. Textile manufacturers began to replace the cottage industry they had always relied on with a vastly more efficient factory system. The result was more goods produced more quickly—and they were cheaper and more readily available, an advance that everyone profited from.

Advances in iron production and Eli Whitney's concept of standard interchangeable parts for manufacturing machinery made the development of machinery more efficient, too. And in 1769 James Watt developed the first practical steam engine. Now, because England had excellent coal and iron ore resources, a cheap source of power was available to streamline almost every industry.

The social effects, however, were not all positive. Factory owners expected long hours of intensive, repetitive work. Working conditions were unpleasant and unsafe, and children, who had always worked alongside their parents in the cottage industries, were employed by the factories and exposed to many hazardous conditions. Workers no longer enjoyed the freedom and independence of working at home, and they spent large portions of every day in unpleasant surroundings, under the watchful eye of a factory boss. Reforms were long in coming and much of the bucolic life of the British countryside was transformed into a series of sprawling, increasingly noisy and dirty industrial towns and cities. But reforms did come eventually, people learned how to use the new technologies more humanely and the Industrial Revolution produced enormous improvements to the general standard of living. In addition, the technologies that it put in motion set the stage for many more advances—including electricity—that would arrive in the following century.

For many historians the period of the Enlightenment ended between 1776 (at the time of the American Declaration of Independence) and 1789 (with the fall of the Bastille and the beginning of the French Revolution). Already, though, before these dramatic events, a reaction had begun slowly to erode the reign of Enlightenment ideals of rational and scientific philosophy. Not everyone had embraced the doctrine of reason. Some thinkers, among them Jean-Jacques Rousseau, detested and feared both science and the newly developing technologies and looked for "enlightenment" through the exploration of feelings and emotions. Still others, including the German author and philosopher Johann Wolfgang von Goethe, composed cautionary books, plays and essays against what they viewed as the arrogance and sterility of the new scientific viewpoint. Feeling that science had robbed both nature and humans of much of their beauty and spiritual values, many people began to turn toward other paths to "truths." Some, like Goethe and the German school of *Naturphilosophie*, preferred to take a more romanticized and "holistic" view of nature, a view sentimentalized and romanticized in England and France by a group of writers and artists called the Romantics. Others, feeling that religion itself was in danger of being corrupted by its accommodation to some of the new scientific methods, turned back to a more traditional approach to the Bible and its teachings.

But, despite this backlash in the arts and religion toward the end of the century, the influence of the scientific method had stimulated productive and positive lines of thought in every area of life, from government to industry, and from psychology and education to philosophy.

PART ONE

THE PHYSICAL SCIENCES IN THE EIGHTEENTH CENTURY

CHAPTER 1

EXPLORING THE NEW SOLAR SYSTEM: THE EARTH'S SHAPE AND THE SUN'S DISTANCE

One of the exciting peculiarities of science is that every new theory produces new questions and casts new light on old ones. The better the theory, the more fruitful the questions it raises. And the theories contained in Newton's *Principia mathematica* were no exception. The French in particular felt uneasy about this force called gravity that Newton claimed was the universal explanation of motion in the cosmos and on Earth. What was the precise nature of gravity? Newton himself could not say. Was it a force inherent in physical objects? To this query, Newton replied, "Pray do not ascribe that notion to me." The French thought this sounded a trifle medieval and referred to Newton's theories as the "metaphysical monster." (We still do not completely understand gravity, although Einstein's description of it as the result of curvatures in space makes it more comprehensible in the 20th century.)

Partly to try to settle some of these questions, early in the 18th century the scientific community—especially the Royal Society in England and the newly formed French Academy of Sciences—became ablaze with two intriguing questions brought into focus by Newton and the scientific revolution: What is the true shape of the Earth? And how far away is the Sun? Both questions prompted a series of exciting expeditions whose scientific inspiration was completely unprecedented in history.

THE BATTLE OF THE BULGE

Newton had predicted, based on his theories of gravity, that the Earth's shape would not be a perfect sphere as the ancient Greeks had thought.

Instead, he proclaimed, it would bulge at the equator and be flattened at the poles by the forces of gravity, such as the pull of the Sun and the Moon. Not only did Newton's prediction present a way to test his theory, but this fairly abstract idea also had very practical implications for people trying to travel in those days—especially those navigating by sea. If Newton was right, all the world's maps were wrong. For the English, the matter seemed urgent: The answers, once found, would not only prove or disprove Newton, they would either transform or reinforce current navigation procedures. The great flurry of voyages and expeditions that followed was rivaled in history only by the search for the Indies and the discovery of America during the age of Columbus and Magellan. In fact, these and other scientific questions inspired such intense interest and so much activity that the historian William H. Goetzmann has called this period the "Second Great Age of Discovery."

Actually the question of the Earth's shape had been in the air for some time—since this was not the first mention Newton made of this idea. So by 1700, the "oblate spheroid" (another way of describing a flattened globe) had become highly controversial, part of an international feud between England and France. On one side was Giovanni Cassini (1625–1712), who was then the director of the Paris Observatory (brought from Italy by Louis XIV). Cassini provided a powerful conservative voice in French astronomy, passing both his conservatism and his position at the observatory on to members of his family for three generations. Cassini not only rejected Nicolaus Copernicus's Sun-centered universe, but also threw his influence behind the cosmic view of René Descartes [day-KART] (1596–1650), a rival to Newton who saw the Earth carried around the Sun in a swirling vortex, in a sort of whirlpool of subtle matter. At rest in the center of the vortex, according to Descartes, the Earth was motionless, as it was carried in orbit by the motion of the vortex around it. To many, his idea made more sense than Newton's "mystery force" that he could not explain, and as a result, in France, followers of Descartes, including Cassini, maintained that the Earth was more the shape of a football—a "prolate spheroid"— elongated at the poles.

In part to prove this point, Jean Richer [ree-SHAY] (1630–96), an associate of Cassini's, set out in 1672 for the town of Cayenne, located very near the equator, in French Guiana on the northern shore of South America. There, coordinating with Cassini in Paris, he performed a series of experiments and observations, among them an exact measurement of the length of a pendulum required to tick off one second of time. To Cassini's amazement, the pendulum in Cayenne was shorter, not longer, than the one in Paris (990 mm. as opposed to 994 mm.). He refused to accept the results and repudiated his former friend. Newton, however, contended that these measurements showed the force of gravity at the equator to be weaker than at the poles, which would be expected if the Earth bulged at the equator.

By this time Descartes's opposing view became a national cause among French scientists and proving it true became a point of honor. Cassini and Jean Picard [pee-KAHR] (1620–82) had led the way by establishing the distance in one degree of latitude, based on the prolate spheroid theory. From that start, they began figuring the path of longitude through Paris to the two "ends of the Earth." And French cartographers were working on the definitive scientific map of their nation.

By the 1730s, though, Newton's theories became popular in France through the writings of Voltaire—whose sharp wit had earned him several years' exile in England—and thanks to translations of Newton's work from Latin into French by Voltaire's friend and lover, Emilie du Châtelet [shah-TLAY]. The French Academy of Sciences decided to settle the question of the shape of the Earth once and for all, by measuring the curvature of the Earth at widely separated locations, as close as possible to the equator and the north pole, where Newton's predictions should be most apparent. Two expeditions were organized, one heading south in 1735 for Peru, and the other departing a year later for Lapland, far to the north.

Led by Pierre de Maupertuis [moh-pair-TWEE], the Lapland expedition included several illustrious scientists, among them Anders Celsius, a noted

Maupertuis and his party measuring the shape of the Earth in Lapland (Figuier: *Vies des savants*, 1870)

VOLTAIRE AND THE POWER OF THE PEN

Born in Paris in 1694, François Marie Arouet, better known by his pen name, Voltaire, was one of the greatest figures of the Enlightenment and one of the most important and stimulating writers in French history. A child of a middle-class family, he attended Catholic Jesuit schools as a boy and later his father forced him to study law. Voltaire, however, had other ideas. He loved writing, freedom and new ideas, and he abandoned law to become a writer. From the first his work was daring and controversial. Some of his earliest works attacked the regent of France and at the age of 22 he was arrested and imprisoned for 11 months.

Instead of frightening and silencing him, Voltaire's imprisonment merely strengthened his decision to speak his mind with all the wit and intelligence at his command. After his release he quickly won fame as a playwright and began moving in the fashionable French literary circles, using his sharp wit and skill to skewer pretensions and injustices that he saw around him. Once released from prison, he spent three years exiled in England, where he quickly made friends with many influential writers and scientists and his agile and discriminating mind found an immediate affinity with the ideas of Newton.

England's political system, which offered more liberty than the French system, was also a tremendous influence on him and he wrote a book explaining Newton's ideas to the French and another entitled *Philosophical Letters on the English* praising England's political ideals.

astronomer from Uppsala, Sweden, and Alexis Claude Clairaut [klar-OH], a brilliant French mathematician, who had published his first book on math at the age of 10 and was admitted into the Academy of Sciences at the age of 18. Maupertuis, a respected French physicist, had visited England in 1728, shortly after Newton's death. There the Royal Society had elected Maupertuis to membership and he had become a great enthusiast of Newton's work, which he promoted eagerly on his return to France. He was delighted with this opportunity to head a scientific journey to prove Newton's theory of universal gravitation and rescue it from the nationalistic controversy in which it had become entangled.

Maupertuis's party set off in 1736 to face numerous hardships in the cold, barren lands of the north, including a near shipwreck in the Baltic Sea. They braved freezing rapids and slept in reindeer skins on the hard, rocky ground, subsisting on wild berries and fish. Hampered by insects and fog, they nonetheless succeeded in completing their measurements and returned,

By the time he returned to France in 1729 his plays and books had made him both rich and famous and he quickly became the most prominent literary figure of his day. Not always loved—his dark, piercing eyes and sharp tongue combined with a mocking manner to make him many enemies—he was nevertheless greatly respected. Addicted to vast quantities of coffee, and totally dedicated to intellectual pursuits, he also set up a small personal laboratory where he conducted physics experiments.

Soon, though, having made more enemies in social, artistic and government circles, he was again forced out of France, and in 1758 he settled down on his estate at Ferney, in Switzerland. From there he conducted what amounted to an international exchange of ideas as he corresponded with most of the leading figures of Europe, including Frederick the Great of Prussia. He also unceasingly continued his battles against the injustices he saw in France. Finally permitted once again to return to Paris in 1778 to attend the opening of one of his plays, he was greeted by vast adoring crowds.

Voltaire died 10 weeks later, at the age of 84. He left behind a legacy of books, plays, essays and correspondence that if not always original in their content (he was much influenced in his thought by Locke and Newton) were always bitingly and scathingly original in their execution. Throughout his life his motto was Ecrasez l'infâme!—"Stamp out abuses!"—and in those words could be found Voltaire's personal code and philosophy: A no-holds-barred battle against bigotry, superstition, dogma, intolerance and anything and everything else that threatened the individual freedom of the human mind and body.

---

triumphant, to France in 1737. Their measurements, published in 1738, showed that the Earth was not a perfect sphere as the Greeks had supposed. Nor was it, for that matter, a prolate spheroid, as Descartes's followers had insisted. Instead it bulged slightly less at the poles than farther south. Maupertuis and his colleagues had proved Newton correct. The success, however, did not win Maupertuis great advancement in his career, perhaps because he had the misfortune to quarrel with the razor-tongued Voltaire, who sarcastically called him le grand aplatisseur—that is, the "great flattener."

Meanwhile, the members of the expedition to Peru had embarked deep upon adventures of their own in the jungles and on the high Andean plains of South America. Led by 34-year-old Charles Marie de la Condamine, an experienced scientist-adventurer, the Peruvian party spent 14 years trekking through steamy jungles and braving the heat and cold of windswept highlands in what has since become Ecuador. Maupertuis had left a year later, swept north, finished the job and returned safely home a dozen years before the return of La Condamine's group. But his measurements were sketchier, less carefully and exactly completed, and for overall impact on 18th-century knowledge, La Condamine's trip outshone his colleague's.

La Condamine, who had studied mathematics and geodesy (the science of the shape and the size of the Earth), had been elected in 1730 to the Academy of Sciences for his work measuring and mapping the coasts of Africa and Asia. Embarking in 1735 from La Rochelle, France, La Condamine and his crew sailed for Colombia and Panama, crossing the Isthmus and proceeding to the port of Manta. There his party split in two, with La Condamine heading 70 miles north with the hydrographer and mathematician Pierre Bouguer (who was of the Cassini school) to survey the first set of measurements along the equator. The rest of the party, which included both French and Spanish scientists, headed for the more southerly port of Guayaquil, the usual launching site for journeys to the town of Quito, located inland on the equator, high in the Andes. Once La Condamine and Bouguer had finished their measurements, Bouguer rejoined the rest of the party heading for Quito, while La Condamine journeyed with surveyor-scientist Pedro Maldonado, governor of Esmeralda, the province in which they were surveying. Traveling by dugout and accompanied by a group of boatmen who had escaped from a slave ship, the two scientists traveled a little-used route up the Esmeralda River to Quito. Surrounded by lush green jungle, hanging vines and diverse, exotic plant and animal life, La Condamine made note of every aspect of his experience. He found himself surrounded by the sounds and sights of a dense, complex world of unimaginable richness: brightly colored toucans, parrots and tiny hummingbirds; chattering monkeys and stalking jaguars; crocodiles and tapirs. He encountered native people who used blow guns, and he took some of their poison back with him to Europe. He watched as jungle dwellers tapped the sap of the caoutchouc tree, noticed them molding the pliable substance into useful objects, and gathered samples of the first rubber to be introduced into Europe. In addition to his studies of mathematics, geodesy and astronomy, he now found himself drawn into the worlds of natural history and anthropology, where his accurate powers of observation and attention to detail served these sciences as well.

Once La Condamine and Maldonado arrived in Quito, the task of measuring the Earth's curvature went slowly. Nervous politicians suspected the team of hunting for Incan treasures and misunderstood the pyramids of stone that they left as markers in their surveys. So La Condamine had to take time out to travel to Lima to gain permission for his team to complete their undertaking in peace. Finally, they succeeded in tracing out a baseline along the high plains. Then they headed south, near Cuenca, where they completed their last measurement in March 1743.

From there, La Condamine once again joined Maldonado to journey across the Andes to the Amazon River, traveling hundreds of miles along its course. La Condamine was the first European ever to introduce thorough observation into the exploration of that territory. Along the way he collected hundreds of botanical samples and continued the observations he had begun along the equator.

La Condamine and his group had taken several years longer to complete their tasks, but their expedition not only measured the shape of the Earth more carefully and accurately than the Lapland party had—providing firm confirmation for Newton—but they established a rich tradition of wide-ranging scientific expeditions, long on perseverance and determination, and even longer on accuracy and encyclopedic fact gathering of every kind.

THE TRANSIT OF VENUS AND CAPTAIN JAMES COOK

The question of the distance from the Earth to the Sun inspired a second wave of expeditions even greater in number and intensity than those investigating the shape of the Earth.

From almost the beginning of time astronomers had tried to measure the distance of the Sun, the Moon and the stars from the Earth. But, because no direct method for measurement exists, the problem proved exceedingly difficult to solve. Two ancient Greeks, Aristarchus and Hipparchus, had tried, without any great success. (Aristarchus figured the Sun to be between 18 and 20 times more distant than the Moon, but in fact the distance is more like 340 times as far. Hipparchus was a little more in the ballpark, but still very far off.) Not until Johannes Kepler (1571–1630) made some key discoveries about planetary orbits some 1,800 years later did anyone come up with a better method, though. Kepler realized that planets orbit the Sun in ellipses and that each planet's average distance from the Sun has a mathematical relationship to the time the planet takes to complete an orbit. So if one could determine the distance of a planet from Earth and observe how long the planet takes to orbit the Sun, then it would be possible to determine the distance to the Sun. Through the use of triangulation, a system of measurement based on trigonometry, it should be theoretically possible to determine the distance from the Earth to a nearby planet. Giovanni Domenico Cassini had tried in 1672 to use Mars for the calculation, using a telescope to measure the small angles. His figure for the distance of the Sun came far closer than anyone else's had before—86 million miles, as opposed to the distance we now know to be 93 million miles—but both the process and the results remained uncertain and frustrating.

Then Edmund Halley pointed out that Venus might be a far better candidate than Mars, since it approaches closer to Earth than Mars. But at its closest approach to Earth, Venus appears so close to the Sun that it cannot be observed at this time except on the rare occasions when it crosses the Sun's disk. This period of crossing, as seen from Earth, is known as a transit, and it is also sometimes referred to as an immersion because, to the observer, the planet seems immersed in the great, glowing globe of the Sun. Halley suggested in 1691 that such a transit would be a perfect opportunity to make measurements, simultaneously, from locations all over the Earth, timing from the moment the immersion begins and the planet appears outlined against the Sun's brilliance until it disappears again on the opposite side. But transits of Venus do not happen often. In fact they only occur, in pairs separated by eight years, at intervals of more than 100 years. The next pair, they knew from orbital calculations, was not due to occur until 1761 and 1769. (Even though he did not live long enough to see the events, Halley was lucky that they were due to occur so soon. The only transits of Venus in the 21st century occur in 2004 and 2012, with none at all taking place in the 20th century.) In 1716 Halley presented a paper to the Royal Society, calling for coordinated, worldwide preparations to begin at once.

And begin they did. The popular press covered the excitement. The best positions for observing Venus at the moment of transit were mapped out. Scientists from all over the world became involved. Several American scientists participated, including the surveyors Charles Mason and Jeremiah Dixon, who traveled to the Cape of Good Hope. In 1761, observers—122 of them—sighted the planet from 62 different locations, from Newfoundland and Siberia, from Peking and Calcutta, from Lisbon and Rome, from the Indian Ocean and from St. Helena off Africa in the South Atlantic. It was perhaps the first great international scientific event, and the enthusiasm was intense, but, despite the care with which the observations were made and the intense preparation, the results were not conclusive.

Because Venus is surrounded by a cloudy atmosphere, its edges appear fuzzy in observation. As a result, an effect known as the "Black Drop" or "Black Thread" kept even the eagerest observer from being able to discern exactly when Venus had traversed completely into the disk of the Sun. Like a raindrop clinging to an umbrella, the outer edge of Venus seemed to cling to the surrounding sky. As a result, measurements made even in the same location by keen observers using identical telescopes turned out to be disappointingly different.

One more chance remained: the 1769 crossing. This time the number of sites was increased to 77 and the number of observers to 151, many in remote, isolated locations, including Baja California, the West Indies, Lapland and the Arctic regions of Russia. Among the most famous was a singular voyage to the newly discovered island of Tahiti in the South Pacific, with the great scientist-explorer Captain James Cook at the helm of a hardy ship named the *Endeavour*.

Captain James Cook exploring New Zealand aboard the Endeavour, *1769*
(The Bettmann Archive)

It was the first of Cook's three great voyages to the Pacific, all three of which gathered a wealth of knowledge about regions previously unknown to the European world. The primary purpose of the *Endeavour*'s expedition of 1768–71, of course, was the observation and measurement of the transit of Venus. Cook himself was an able astronomer, whose ability to navigate by the stars was prodigious. He was joined in the observation of Venus by astronomer Charles Green. In addition, the scientific team included Sir Joseph Banks, a wealthy young explorer-artist, who provided the finances to purchase scientific supplies and equipment for the eight other members of the *Endeavour*'s international natural history corps. Several prominent Scandinavian scientists were included, one of whom was a protégé of Carolus Linnaeus [lih-NAY-uhs], whose classification system of plants and animals, as we will see in Part Two of this book, had by 1753 already revolutionized the way people organized their thinking about living organisms.

The ship arrived in Tahiti in April 1769, allowing plenty of time to build an observatory (at a spot still known as Point Venus) before the moment of transit on June 3. But the day of the transit proved disappointing. As Cook wrote in his journal:

> *Saturday 3rd. This day prov'd as favorable to our purpose as we could wish. Not a Clowd was to be seen the whole day and the Air was perfectly clear, so that we had every advantage we could desire in Observing the whole of the passage of the Planet Venus over the Suns disk: we very distinctly saw an Atmosphere or dusky shade round the body of the planet which very much disturbed the times of the Contacts particularly the two internal ones. Dr. Solander observed as well as Mr. Green and my self, and we differed from one another in observing the times of the Contacts more than could be expected. . . .*

The conditions seemed perfect, but even so, the instruments were not powerful enough to provide definitive, accurate measurement.

Once all the measurements made of the transit of Venus worldwide in 1769 were compiled and analyzed—and it took nearly 60 years—the average value achieved turned out to be 96 million miles, actually much closer to the correct 93 million miles than any measurement made before. (The reading we accept today, achieved with an accuracy of several decimal places by radar measurement techniques, was not reached until the mid-20th century.) Based on the measurements made in 1769, the Solar System had turned out to be nearly 100 times the size that Ptolemy had estimated the entire universe to be—a staggering change in world view. But the scientists of the 18th century were not satisfied; they longed for more exacting measurement, and in this aspect of Cook's voyage he was clearly disappointed.

Cook's other main objective in 1769 was to sail south to look for the great lost continent of Terra Australis Incognita. In this, too, he failed. But, as often happens in science, in his failure also lay his great achievement. For he sailed in waters never before explored. He discovered 70 islands he named the Society Islands, touched ashore in New Zealand, impaled his ship on the Barrier Reef, discovered Botany Bay and Cook's Bay. His naturalists saw kangaroos for the first time and Banks and his colleague, Daniel Solander, collected 17,000 new species of plants, as well as hundreds of species of fish and birds and skins of many animals. The artists aboard returned to Europe with drawings, sketches and paintings of exotic places, native customs and plants and animals the likes of which the people at home had never imagined. It was the first of three voyages Cook would make, the last ending in tragedy, when he was killed by Hawaiians who only days before had worshiped him as a god. All three expeditions returned, however, with enormous discoveries about the peoples, flora, fauna and landmasses of a world previously unexplored.

James Cook epitomized the kind of scientific eclecticism that prevailed in the 18th century, when a great navigator could contribute significantly to the science of astronomy with the precision of his measurements and observations, to geography with his maps and charts, and to botany, zoology and anthropology with his copious observations and comparative descriptions. It was a time when great minds could contribute in many fields. In fact, it was a time when those fields were not so clearly distinguished from each other as they would later become—a time when chemists were both physicists and physiologists, when geologists were also botanists and zoologists, and when mathematicians and musicians were also astronomers.

CHAPTER 2

OBSERVING DEEP SPACE: STARS, GALAXIES AND NEBULAS

For those gazing into deep space and trying to unravel its mysteries, the 18th century had dawned with a brand-new set of challenges. Up until the end of the 17th century, throughout time, the complex paths of the planets had absorbed most of the attention of positional astronomers. Records dating back thousands of years B.C. show that the earliest stargazers watched these "wanderers" and based complex philosophies and occult practices such as astrology on what they saw. The stars, by contrast, appeared to be more predictable, fixed in the skies, except for movements now explained by the revolution of the Earth around the Sun. The ancients—even the savvy ancient Greeks—believed that the stars, unlike the planets, were unmoving. The great Greek philosopher Aristotle and his follower Ptolemy had envisioned a universe composed of nested spheres, layered like an onion around the Earth, with the motionless stars embedded in the outer layer. The new 17th-century Copernican vision of the heavens straightened out a lot of the confusion about the Solar System, recognizing the position of the Sun at the center. But even this new, revolutionary view of the universe placed the stars motionless beyond the edge of the planetary orbits. Copernicus, Kepler and Newton, with his laws, had explained the motions of the planets. But if Newton's laws were truly "universal," then did the stars also have predictable movement? Newton did not address this question. For that, we turn to a young disciple of Newton's, Edmund Halley.

OF STARS AND COMETS

Edmund Halley was born in November 1656, the son of a wealthy London soap manufacturer. By the time he became a student at Oxford he had

THE REMARKABLE JOHN GOODRICKE

In the brief 21 years of his lifetime, astronomy offered John Goodricke both consolation and fame. Born in 1764 in the Netherlands, the son of an English diplomat, Goodricke spent his brief life in a world of silence, unable to hear or speak. His parents, though, were both wealthy and understanding and encouraged his interest in astronomy in his early teens. Under the night sky, sparkling with stars, Goodricke seemed to feel at home with his silence. Gifted with a keen mind and an observant eye, he turned each night to his telescope.

One star in particular caught his attention. Algol, in the constellation of Perseus, was a source of puzzlement for astronomers. Aristotle had taught that the stars were perfect and unchanging, but Algol, shining brightly in the sky, appeared to display a variability in its brightness. Perhaps it was this uniqueness that drew the young Goodricke to the star. Like him Algol was an anomaly, standing alone in its difference from the stars around it.

The young astronomer took it upon himself to attempt to decipher the mystery of Algol, a job that called for excruciating patience and dedicated observation. Much of the work was done with the naked eye (that is, without the aid of a telescope), since he needed to make comparisons to stars beyond the field of vision of a telescope.

developed an interest in astronomy, and at 19, while still a student, he published a work on Kepler's laws. The little book caught the attention of John Flamsteed, who became Britain's first astronomer royal that same year and had already begun putting together a new catalog of the stars in the Northern Hemisphere. Flamsteed, impressed with Halley's work, encouraged the young man to go further.

By the mid-1670s, the improvement in telescopes and instruments had revealed many inaccuracies in the old star charts and catalogs. New ones were badly needed. So, while Flamsteed and others were busy charting the northern skies, it occurred to Halley that the skies in the Southern Hemisphere were open territory to an ambitious astronomer. He convinced his father to finance his trip, and, taking a leave from his studies at Oxford, Halley boarded a ship to chart the stars hanging over the Southern Hemisphere.

For the next year and a half, from 1676–78, Edmund Halley spent hour after lonely hour with little more than his telescope to keep him company on the bleak island of St. Helena, off the southwest coast of Africa. The

Months of intense study and careful note taking convinced Goodricke that the star's variations followed a regular pattern, a specific period or cycle. After much more intense study he was able to demonstrate that Algol went through its light changes in a regular period of just under three days. It was the first demonstrable proof of the regularity of a variable star. Today we know that more than 29,000 stars are "variable" and more than 14,000 other light sources appear to exhibit variation in their light.

In 1782, Goodricke attempted to answer the difficult question about the source of Algol's variation by suggesting that it was due either to dark spots on its surface or to an invisible (to us) companion that orbits the star and comes between us and the star's light at regular intervals. It was a daring hypothesis, especially when put forward by an unknown 17-year-old amateur astronomer. And, although Goodricke's theory about Algol's mysterious orbiting companion would not be proven until much later—when it was discovered that Algol had not just one but two companions—his paper won him the coveted Copley Medal from the Royal Society. Goodricke's death in 1786 came just before his 21st birthday, only two weeks after he had been elected a fellow of the Royal Society. The young man had always suffered from ill health, and his death was attributed to overexposure to the cold night air.

Of the more than 29,000 variable stars known today, few are as easily explained as was the mystery of Algol, and Algol itself still contains many anomalies. Thanks to young Goodricke, though, another small chunk of knowledge was added to our understanding of the heavens.

weather was harsh and inhospitable, extremely bad for astronomical observation and no kinder to young Halley. But finally, his stay completed, Halley returned to England with a new and unprecedented catalog of the southern skies, listing the positions of no fewer than 341 previously uncharted stars. Halley had hoped to chart many more had the weather been more favorable. But the work, published as *Catalogus Stellarum australium* ("Catalog of the Southern Stars") in 1679, was enough to make Halley an immediate celebrity among London's scientific elite. Flamsteed heralded Halley as "the southern Tycho" (referring to the ground-breaking work of the great Danish astronomer, Tycho Brahe). Years later, Flamsteed would quarrel with Halley over the premature publication of one of Flamsteed's own star catalogs. But this was Halley's hour, and, gentlemanly and charming, he made the most of it. A tremendous success, Halley was invited to join the prestigious Royal Society. It was heady company that young Halley suddenly found himself surrounded by, including such giants as Newton, Robert Hooke, Flamsteed and the charismatic Christopher Wren. But intelligent, modest and personable, Halley found quick acceptance. Surprisingly he even struck up an immediate friendship with the reticent and usually sour-faced Isaac Newton. Among the first to see the *Principia*, Halley not only encouraged Newton to publish his revolutionary masterwork but paid for the publication out of his own pocket when funds in the Royal Society ran short. Money had never been a problem to Halley, and when his father was found mysteriously murdered, Halley's inheritance guaranteed him a comfortable income for life.

The leisurely life held little appeal for the intellectually curious Halley, however, and he was soon once again hard at work. During his long voyage south to St. Helena he had observed that the ship's compass did not point exactly to the north pole as was commonly believed. Thus the magnetic pole and the north pole were not the same. And although the difference might be construed as minor, for navigators and ship's captains the discovery was an important one. The worldwide explosion of commercial trade during the later part of the 17th century had opened up many new ocean routes and competition was fierce. Just as new maps of the skies were needed, so too were new and better marine charts for use by seafarers for efficient navigation. In 1698 under the command of Halley the first ship ever commissioned and outfitted solely for the purpose of scientific duties sailed out from London. Christened the *Paramour Pink*, the small but sturdy ship toured the seas for two years while Halley and his crew measured magnetic declinations around the world, composed new navigational charts and attempted to determine the precise latitudes and longitudes for dozens of major ports.

Returning to London a seasoned sailor, Halley earnestly began another kind of research, this time delving through hundreds of stacks of ancient astronomy books and charts. During his journeys Halley had become intrigued with the many discrepancies between the ancient star charts and more contemporary ones. Despite the numerous outright and often outrageous errors in the old charts, he had noticed that many correctly charted stars still differed in their positions from modern observations. After careful study he was able to prove that at least three major stars, Sirius, Procyon and Arcturus, had changed their position since they had been charted by the Greeks.

What some astronomers were beginning to suspect, Halley proved to be true. The stars too were in motion. Even beyond the Solar System the objects of the universe appeared to be in constant dynamic and predictable movement—an idea inherent in Newton's laws of gravitation but unproved until Halley's careful work. Once again the ancient views of the universe were shattered under the power of Newton's grand gravitational and geometrical vision. This new piece of information would have enormous impact on astronomy in the years to follow.

But Halley is best known for his work with comets, long held in awe and superstitious reverence for their strange appearances in the night skies. As

Edmund Halley began a long tradition of curiosity about the comet named for him. In 1985 the European Space Agency launched this spacecraft named Giotto *to observe Halley's Comet during its 1986 approach to the Sun (an artist's conception of the encounter at a distance of 376 miles, with the painting of the comet based on the Mt. Wilson Observatory photograph taken of Halley's Comet on May 8, 1910).* (European Space Agency)

Halley pored over the ancient records, he discovered that he could see a pattern that fit Newtonian theory when he attempted to calculate the shape of their orbits, based on records of their appearances over the centuries. Precise information was sparse, however, and he was forced to abandon his project, but not until after he had made an interesting discovery. His calculations strongly suggested that four well-known comets, seen in 1456, 1531, 1607 and 1682 respectively, were in all likelihood just one comet, which, following Newton's gravitational laws, returned on its orbit once every 76 years. It would return, Halley predicted, in 1758. And, although he died in 1742, 16 years before he could see his prediction come true, the comet, known today as Halley's Comet, remains a dramatic tribute to the memory of the astronomer who first recognized the regularity of its movements.

NEW VIEWS OF THE UNIVERSE: WRIGHT, KANT AND LAPLACE

If the stars as well as the planets were in motion, then the universe beyond the Solar System was no longer necessarily the sacrosanct, static and perfect place it had always been thought to be. And, if Newton's laws held true for all objects throughout the heavens, as they were appearing to by the mid-18th century, then new perspectives and speculations about the universe in general were called for.

In the mid-1700s a handful of men put forward theories that began to revolutionize humankind's view of the cosmos and our own position in it. One of these was the French writer and naturalist Georges Louis Leclerc, comte de Buffon, who will be discussed later in this book. Among the others were a deeply religious amateur scientist, a young speculator in philosophical ideas and a brilliant mathematician.

Thomas Wright

Thomas Wright (1711–86) was an English thinker of many diverse interests. Well educated and well read in the prevailing philosophical and scientific views of his day, like many thinkers during the 1700s he was troubled by the conflicts between his scientific observations and thinking and his religious beliefs. His astronomical observations and reading had led him to speculate that if the planets moving around the Sun formed an orbital system, then, given Newton's laws of gravity, it was possible that the stars themselves might also be a part of one or many similar orbital systems. If the heavens

The galactic center of our own Milky Way Galaxy as seen through a 20th-century telescope (the streak of light extending from the upper left corner is the trail of an early artificial satellite orbiting the Earth). In 1750 Thomas Wright daringly speculated that our Sun was just one of many stars orbiting the stellar system's "divine center." (U.S. Naval Observatory)

were thus composed, though, where was the domain of God in this new mechanistic and gravitationally interacting universe? Neither a trained astronomer nor a rigorous and systematic thinker, Wright actually proposed a number of different cosmological models. One of the earliest pictured the Sun and Solar System, moving along with the stars to circle a huge common center, which might or might not be a solid, but nevertheless was the domain of God. On the outer region of this gigantic system, far beyond the Solar System and stars, was a mysterious dark region of "doom" that surrounded all. It was a highly mystical concept but further elaborations proved more fruitful. In his book *An Original Theory and New Hypothesis of the Universe*, published in 1750, Wright suggested an alternative possibility: that rather than a shell-shaped universe in which the stars and Solar System moved around a common center, the entire system might actually take the form of a flattened rotating disk. He still clung to the idea of a common center occupied by God or a divine presence, but he also suggested that the so-called nebulas, the mysterious murky white patches in the sky that were neither stars nor planets, lay far beyond the disk itself.

Immanuel Kant

A deeper and more rigorous thinker than Wright, the young German philosopher Immanuel Kant [KAHNT] (1724–1804) had read either Wright's provocative book, or a review of it. In any event he quickly turned his razor-sharp mind to some of Wright's loosely formulated ideas.

In 1755 he offered his thoughts in his book *Allgemeine Naturgeschichte und Theorie des Himmels* (*General Natural History and Theory of the Heavens*). As a work of speculative cosmology and astronomy it was a truly remarkable achievement. Following Wright's lead and his own careful studies in Newton's physics, Kant suggested that the stellar system, like the Solar System, was shaped like a flat disk. Just as the planets in the Solar System revolve around the Sun in elliptical orbits in the plane of the zodiac, so, suggested Kant, the stars might also revolve around a common and unknown center. Unlike Wright, Kant made no reference to a divinity occupying this hypothetical central position. Kant also suggested that the visible stars made up this common "galaxy," but that the system in which our Sun and planets joined those stars in common revolution around an unknown center was probably not the only such system to exist in the universe. The universe, in fact, might be composed of many such galaxies. Kant further suggested that we may already have seen those other galaxies, that they might in fact be the nebulas observed by astronomers and mentioned by Wright. Many of these nebulas, Kant pointed out, were oval- or disk-shaped and the combined light of the stars making them up might, at such a great distance away from us, create their milky and blurred appearance. It was an inspired and

CHARLES MESSIER AND HIS CATALOGUE

Charles Messier (1730–1817) was a comet hunter. Like many astronomers in his day and ours, he was fascinated with finding and tracking those elusive and dramatic visitors in the night skies. During his lifetime he is credited with discovering more than 15 new comets, earning him the nickname of Comet Ferret. Like most dedicated specialists, however, Messier was somewhat fanatic about his field of specialty and did not care to be disturbed with irrelevancies. And, in the early days of his telescopic investigations, he found that he was constantly distracted by those small milky-white objects in the sky that were called nebulas (the Latin word for cloud, which is what they look like). Not much was known about them when Messier began his work. Immanuel Kant had made a few speculations, but William Herschel's classic work was yet to come, and anyway, Messier was after comets and couldn't care less about nebulas. His problem was simply that those nebulas could easily be mistaken for comets, moving very slowly in the distance, and a great deal of time could be wasted watching them to make sure that they were not comets but permanent fixtures in the night skies.

Messier decided that the only way to handle the problem once and for all was to compile a catalogue, or list, of all the distracting nebulas and their exact positions in the sky so that other comet hunters would know exactly where they were located and would not waste valuable time on them.

It was a tough job, given the still primitive nature of the telescopes of his time, and it demanded a great deal of patience. Messier, however, was able to catalog 103 of these offending objects by 1781. Today, many amateur as well as professional astronomers, wishing to view such wonders as the great Andromeda Galaxy (M31 in Messier's catalogue—he preceded each entry with M, for Messier) and such other marvels as globular clusters, open clusters, planetary nebulas and even a supernova remnant, can find them quickly thanks to Messier's famous work.

daring speculative leap and one, although not proven until the 20th century, that opened up a vast new horizon for cosmology: an immense oceanlike universe filled with floating, rotating, island galaxies—a universe much larger and grander than anything any human could ever have dreamed.

It is important to remember here, though, that Kant's ideas, although based on his understanding of Newton's physics, were, however inspired, purely speculative. Although he was well read in physics and mathematics, Kant, who would later gain much fame and controversy as a philosopher, was neither a working physicist nor a mathematician.

Pierre Simon Laplace

Not so Pierre Simon Laplace, born in Normandy, France in 1749. Laplace was a brilliant and inspired mathematician, one of the most influential scientists of the 18th century. Although he is largely unknown by most people today, his rigorous mathematical work nailed down some of the loose ends left by Newton and provided a stronger foundation for the speculation of Kant.

In 1773 he made his first major contribution to astronomy when he examined the orbit of Jupiter in an attempt to understand why it appeared to be continuously shrinking while the orbit of Saturn appeared to be expanding. In a three-part paper published between 1784 and 1786, he demonstrated that the phenomenon was periodical—that is, occurred at regular intervals—and had a period of 929 years. While carefully explaining the reasons for the phenomenon, Laplace's paper also served to solve a major problem left open by Newton, that of the stability of the Solar System. Newton, disturbed by the complex gravitational interactions between the

Pierre Simon Laplace
(The Royal Society, England)

bodies of the Solar System, had concluded that some kind of divine intervention was needed occasionally to keep the entire system stable. Aided by fellow mathematician Joseph Louis, comte Lagrange (1736–1813), Laplace demonstrated mathematically that since all of the planets orbit the Sun in the same direction, their eccentricities and inclinations in relation to one another always remained small enough to need no outside intervention to remain stable in the long term. Or, as Laplace wrote:

> *From the sole consideration that the motions of the planets and satellites are performed in orbits nearly circular, in the same direction, and in planes which are inconsiderably inclined to each other, the system will oscillate about a mean state, from which it will deviate by very small quantities.*

In effect, the Solar System, said Laplace, was essentially a self-correcting mechanism, and it needed no divine force to nudge it back in place.

Laplace wrote many other important papers, shedding much mathematical light on several other large and small problems related to Newton's theories and the gravitational interactions of the planets and satellites. His most popular and widely read work, however, came in 1796 when he

Joseph Louis Lagrange (Courtesy, Burndy Library)

published *Exposition du Système du Monde*, a popular work on astronomy, and in it, almost as an afterthought, offered a theory to explain the origin of the Solar System.

The "nebular hypothesis," as Laplace's idea came to be called, had also been offered, perhaps unknown to Laplace, in a less rigorous and slightly differing form by Kant. Variously called the nebula hypothesis, Kant's nebular hypothesis, Laplace's nebular hypothesis and the Kant-Laplace nebular hypothesis, the idea became immediately popular and was accepted by most astronomers throughout the 19th century.

In a much shortened form, and stripped of its mathematical framework, the idea was that in the beginning the Sun originated as a giant rotating nebula or cloud of gas. As the nebula rotated and the gas contracted, the speed of its rotation increased until the outermost material in the nebula was moving too quickly to be held by its gravitational force. Stripped away, this gaseous material would then condense to form a planet, while the central nebula continued to increase its spin and contraction, eventually leaving behind more loosened material to form another planet, and so on, until all the planets were formed and the central nebula formed a stable Sun at the center. The theory was fraught with many problems and unanswered questions, but as an early attempt to explain the origin of the Solar System using scientific and rational reasoning, without recourse to occult or metaphysical forces, it was an impressive beginning.

Although Laplace himself knew that he had crossed the bounds into speculation with much of this "theory," he had no doubt that this and all other mysteries of the heavens would be resolved by the reasoned applications of mathematics and Newton's laws. Not the most modest of men nor the most gracious of scientists (he was often both hated and admired by his associates), he was one of the Enlightenment's most devout believers in the mechanistic nature of the universe. It was a belief held by many scientists during the Age of Reason, but few stated it as forcefully as Laplace when he wrote that in Newton's universe nothing happened by chance and that, in a hypothetical case, "an intelligence knowing, at a given instance of time, all forces acting in nature, as well as the momentary position of all things of which the universe consists, would be able to comprehend the motions of the largest bodies of the world . . . [and] nothing would be uncertain, both past and future would be present" in such a person's eyes.

Not all of Newton's followers during the 18th century were such ambitious philosophers though. While theory still had its place, many astronomers, particularly in England, shied away from such grand theoretical speculations, preferring to take a more pragmatic and practical approach by employing the lessons of the scientific revolution and Newton's laws to everyday working astronomy.

William Herschel, one of the greatest observational astronomers of the 18th century
(Courtesy, Burndy Library)

TEAMWORK: WILLIAM AND CAROLINE HERSCHEL

If the mathematician Laplace was the most influential theorist in astronomy during the 18th century, then the musician and amateur astronomer William Herschel [HUHR-shuhl] was the most influential working astronomer of the Enlightenment. Born in Hanover, Germany in 1738, Herschel fled from military service in 1757 and moved to England. A renowned musician in Germany (oboist, organ master, violinist and concert master), he picked up his career in England, teaching and copying music. In a short while, though, he was well established and began to find positions as a conductor of military bands and as a respected composer and organist. In 1766 he was hired to become organist at the Octagon Chapel in Bath. It was a good position, allowing him time to teach, conduct and compose. More important, it paid well and supplied him with enough money to help finance his developing interest in astronomy. By the time his sister, Caroline (1750–

1848), joined him in England in 1772, William had become so enamored of astronomy that he had devoured hundreds of books on astronomy, calculus and optics. He had also purchased a small telescope and had begun to spend his nights gazing at the skies above Bath.

Caroline was quickly caught up in her brother's enthusiasms. An accomplished musician and singer, she was soon helping William with his orchestrations, copying his music, and displaying an almost equal fascination with astronomy and telescopes. It was only a short time after his sister's arrival when William, with Caroline's help, constructed his first small telescope. From that moment on there was no turning back. Another, slightly larger telescope followed quickly, and then another.

And interesting telescopes they were, much wider than the usual telescopes of similar length. When asked about their unusual structure Herschel would explain that he was after not only magnification but light-gathering ability. The larger the telescope's mirror, the more light it was capable of

Caroline Herschel, the first woman to play a key role in astronomy, enthusiastically spent long hours with her brother, William, "sweeping" the nighttime skies. As Caroline wrote in her diary, "If it had not been for the intervention of a cloudy or moonlit night, I know not when he or I either would have got any sleep." (The Bettmann Archive)

capturing, and the more light it captured, the more stars and nebulas he could see. Herschel had already become fascinated with the universe beyond the Solar System. Over the long progress of his career he would, with Caroline's help, define the discipline of stellar astronomy and focus the eyes of astronomers far beyond the mechanics of the Solar System. Curiously, though, it would be a discovery within the Solar System itself that would bring him his greatest fame, energize observational astronomy and make him a worldwide celebrity.

Although Herschel could be a patient and precise observer (he once measured more than 100 different lunar craters, triple checking himself by using three different methods), he was much more interested in discovery than in calculation. The night sky for Herschel was a vast, dark ocean, poorly charted and filled with the perpetual promise of new discoveries. To aid him in his nocturnal navigation through the heavens he had devised a careful method of "sweeping" the skies. Each night he would work on only one small area of sky, usually a strip of only about two degrees, which he would explore twice before retiring. The next night he would examine an adjoining strip, gradually working around the entire night sky visible from his location.

It was on one of his careful sweeps in March of 1781 that Herschel made his famous discovery of a new planet far beyond Saturn (the most distant planet known to astronomers at the time). Other astronomers had seen the planet, which soon was named Uranus, but everyone had always thought it was a star and quickly passed by. During his nightly sweeps, though, Herschel had memorized the positions of thousands of stars and he quickly recognized the small speck of light as a "wanderer," in a place where no star ought to be. Was it a comet or a planet? Reporting his discovery to the astronomer royal of the time, Herschel waited for his answer. A planet, reported the astronomer royal after calculating its orbit, definitely a planet, and so far away that it changed the size of the known Solar System to nearly double what everyone had thought. William Herschel was suddenly famous.

Curiously, after Newton astronomy had become a pretty tame business in most everyone's eyes. Newton, after all, had explained everything and just about everything that could be known was known. The theorists were working up their armchair philosophies, but most of them were pretty difficult to take seriously (if the average person ever heard about them at all), and the working astronomers were busy calculating and cataloguing. New stars were occasionally announced, but for most people, well, a star was a star. A new planet though, now that was something different, something unknown before and closer to home, something that the average person could actually imagine. And, even more exciting, maybe there were others yet to be discovered in the Solar System. Maybe the scientists did not know everything there was to know after all!

This musician and amateur astronomer had suddenly made astronomy exciting again. The excitement was almost too much for Herschel, though. Appointed a fellow in the Royal Society, he was wined, dined, feted and appointed court astronomer to King George III. This last honor had a major drawback, as Caroline quickly realized. With it went a pension, which was to enable her brother to quit his job as a musician and become a full-time astronomer. Unfortunately, the pension was considerably less than his earnings as a musician. As William pointed out, however, one did not turn down an appointment from a king. Other means would just have to be found to keep the money coming in. They were accomplished telescope builders, in fact some had said that their telescopes were the best in the world. The brother and sister would build and sell their fine instruments for additional income.

In the meantime, with more time on his hands to give to his new career, William could devote himself to building those wondrous, gigantic instruments he had always dreamed about, instruments that would enable him to look deep into space. "I have looked farther into space than ever [a] human being did before me," he wrote. Uranus had been a feather in his cap. In 1787 he also discovered Titania and Oberon, two of Uranus's moons, and, before he was finished, two more moons around Saturn. But Herschel's first love remained the starry realm and the great telescopes that enabled him to peer into its depths.

Not all of his telescopes were successes. One "monster" crashed under its own weight as Herschel was climbing to its mount. Another, planned around a mirror three feet in diameter, had to use an "enourmous amount of horse manure" (sifted and pulverized by the ever-enthusiastic Caroline) for the mirror's mold. The day of casting, the overheated mold cracked, sending molten metal cascading across the floor and turning the place into a miniature, foul-smelling Vesuvius, while Caroline and William took to their heels in frantic escape.

While it must have seemed to some people that the Herschels were spending all of their time building telescopes, they managed to carve out more than enough time for their observations. And, when William married a wealthy woman in 1788, they were able to stop building telescopes for sale and concentrate on their own instruments and observing.

Edmund Halley had proven that the stars were not "fixed" but had a "proper motion" in the skies. In 1783, Herschel made his first major "stellar discovery" when his observations proved that the Sun too was in motion across space. Our own Solar System was moving in its entirety. The Sun, planets and moons not only performed their intricate dance among themselves, but the dance floor itself was in motion as the entire system swept through the heavens. Not only was the Sun no longer the stationary center of the universe, but now, it appeared, neither was the Solar System. Once more, humankind found that it held no privileged position in the universe, but inhabited a globe that was but one small participant in a great cosmic dance. And more and more it became apparent to the 18th-century thinkers that the entire universe was an endless dance of movement.

Not only was everything in motion, but there was more of "everything" than anyone could have imagined.

William Herschel had always been drawn to the mysterious patchy nebulas in the skies and to the conjectures by some thinkers that they were vast collections of stars too distant to be clearly seen. With his larger and more accurate telescopes he was able to resolve many of them, proving once and for all that many were vast collections of stars, or galaxies. His early conclusion that all nebulas were vast galaxies of stars was later proven to be incorrect, as many nebulas refused to resolve into a clear demonstration of stars, but many of these were, he suspected, vast clouds of some "shining fluid" that was "the chaotic material of future suns."

In a series of long observations of so-called double stars made with Caroline during the 1780s he also proved that Newton's laws were indeed

Caroline Herschel, who independently discovered eight comets, lived to the age of 98 and devoted her entire life to astronomy. (Courtesy, Burndy Library)

universal. His careful observations proved that many of the double stars were not just pairs, but so relatively close to one another that they were locked by gravity into a mutual orbital dance. It was still further proof that Newton's laws stretched far beyond the Solar System, throughout the entire vastness of space.

William Herschel was knighted in 1816, continuing his observations nearly up until the time of his death at the age of 84 in 1822. More than any other astronomer of the Enlightenment, Herschel brought astronomy into the modern age. His telescopic observations gave reality to the speculations of those thinkers during the Age of Reason who were beginning to see and understand humankind's place in the universe—that we on this humble planet called Earth were but one small part of a vast cosmos swimming with stars and galaxies.

For some this knowledge was frightening. Humankind and the Earth seemed to become more and more insignificant in the eyes of science, no longer the unique privileged and special creations of God. For others, though, a wonderful, new world of knowledge was opening up. As Halley, Herschel and other observational astronomers began to show what was, thinkers like Kant, Laplace and others attempted to understand how, given the apparently mechanistic nature of the universe, it all came to be. How did all the pieces fit together to form a whole? If the universe was filled with galaxies, by what process did the galaxies form? How did stars and planets form? How did the universe itself come to be? Did it have a beginning; did it have an end? They were big questions that would command many years even to develop serious theories about. Much more needed to be learned, and many new instruments and techniques developed. Much more basic physics remained to be understood. What was this mysterious "gravity" that "controlled" so much? What was light? What was heat? How did the stars "shine"?

It was a demanding and complex agenda—one that would occupy many scientists for centuries. Not since the ancient Greeks had humans taken it upon themselves to attempt to explain so much. Newton though, after all, had given them the key. It was both the glory and the hubris of the 18th-century thinkers that they believed that with it they could quickly unlock all of nature's mysteries.

CHAPTER 3

REDISCOVERY OF THE EARTH: ROCKS AND AGES AND THE BIRTH OF A NEW GEOLOGY

As 18th-century astronomers continued to focus on the deep reaches of the cosmos and the fascinating realm of the Solar System, another group of thinkers began to turn their attention to the history of the Earth itself—and its origin. If the heavens could be understood as a vast clockwork mechanism operating under Newtonian laws, was there also a mechanism that explained the creation of the diverse features of the Earth? Why were there lands and seas? Mountains and valleys? Different kinds of rocks? What were those strange features in some rocks that looked like embedded creatures or objects?

The ancient Greek philosophers had speculated on ways in which many of the Earth's features might have been formed. Thales [THAY-leez] of Miletus (c. 624–c. 546 B.C.), had observed the violent power of waves breaking against the shore, and the less dramatic but no less persuasive power of the steady erosion of river beds by meandering rivers. It was quite obvious, said Thales, that water was tremendously important in its ability to change the features of the Earth. Water in fact was so important to Thales' thinking that he believed it to be the source of everything on the Earth, both animate (living) and inanimate (nonliving). The Earth itself, said Thales, could be seen as a ship afloat in a gigantic ocean, and waves or disturbances in that ocean might be the causes of earthquakes as our planet bobbed in its surrounding environment. Although many of the ideas of Thales and the other ancient Greeks may seem incredibly naive to us today, his recognition of the power of water as a fundamental force that helped to sculpt and form the surface of the Earth was an important early observation. Thales' speculations, of course, were just that—speculations. The great

Greek thinkers were philosophers, not scientists. And since the ancient Greeks had no desire and made no attempts to scientifically "prove" their ideas, there was no way that one speculation could be taken to be more correct than any other. Everyone had a right to an opinion, provided that their intellectual credentials were in order. As long as their ideas had internal consistency (that is, if all parts of their systems appeared to fit together without any one part contradicting another), then it was simply a matter of paying your money and taking your choice.

By the time the intellectual, commercial and political power of the world shifted to Rome in about 100 B.C. the Romans had abandoned most such speculations altogether. More practical than philosophic, the ancient Romans preferred to spend their time figuring out the best way to build roads, bridges, tunnels and aqueducts to expand the power and prestige of the Roman world. For the Romans it was a fruitless exercise to worry about how the mountains got there and much more important and challenging to figure out how to put passable roads across or around them.

After the fall of Rome, when most intellectual activity in the Western world was confined to monasteries, much of the light of learning shifted to the East. Even the Arabs, however, who had borrowed, preserved and expanded on much Greek learning, displayed little interest in what we today call geology. The great Arab philosopher Avicenna (A.D. 980–1037) did attempt a rough-and-ready system, borrowing from and adding to some of the thoughts of the ancient Greeks, but there were still many problems with most of his explanations.

As the light of learning began to filter from the East slowly back into the Western world during the Renaissance (beginning mid-14th century), the study of geology remained low on the list of intellectual priorities. The brilliant artist Leonardo da Vinci, who took an interest in just about everything, put forth his ideas on the subject around 1508. Echoing some of Avicenna's thoughts, including the idea that mountains were created by the erosion of high lands, Leonardo also suggested that rivers were fed by rain and melting snow and that the intriguing and mysterious fossil shells that had been found in northern Italy were the remains of once living marine creatures that had died when the land was beneath saltwater. For the most part, though, Leonardo's thoughts on geology made little advance on Arab or ancient Greek thought.

The picture was further complicated by the presence of the Christian church, which almost single-handedly kept the flickering light of civilization alive through the Middle Ages, but had in the process grown into a powerful, single-minded and authoritarian force. While the ancients had demonstrated only a passing interest in the problems of geology, they had not run into any formal or established opposition when they ventured into speculations on the subject. If they remained relatively silent or lacked profundity, their lack of progress was caused more by their own lack of interest or discipline than by interference by established authority.

With the increasing power of the Christian church, though, much had changed. In the 17th century Galileo had provoked the wrath of the church when his astronomical observations led him to the conclusion that the Sun and not the Earth was the center of the Solar System. The idea was in direct conflict with established church doctrine based on the word of the Bible and the astronomical system handed down by Ptolemy. For the church the revealed word of God was sacred and the Bible was the final authority on all matters. Christians, seeking to understand the history of the Earth, had merely to turn to the Bible for their answers. There in the stories of Genesis and the great catastrophe of Noah's Flood was the history of the Earth encapsulated.

The age of the Earth worked out in 1654 by the Irish prelate James Ussher (1581–1656), carefully calculated according to biblical chronology, set the Earth's first day as October 26 of the year 4004 B.C., probably at 9:00 A.M. The entire creation, as the Bible stated, had taken a week. Ussher's calculations were well received at the time, since most theologians were agreed that the Earth could not be less than 5,000 or more than 6,000 years old. It was against this background that the early students of the Earth had to proceed.

One of the problems, of course, as any careful observer could see, was that the Earth appeared much older than the few thousand years allotted for its existence. Everywhere one looked at its broken and weathered surface, its giant mountains, rugged cliffs and deep valleys, there was evidence of a planet that had suffered a great deal and gone through many changes. In fact, it was generally agreed by theologians that the Earth was in serious decline and was probably not going to last much longer. How then had it changed so much in such a short period of time?

Another problem was the presence of fossils embedded inside rocks. Although they had been known since the ancient Greeks, their existence had never ceased to be troubling. By the closing years of the 17th century many serious thinkers had come to accept fossils as the organic remains of once living creatures. But, as we will see later in this book, many others still held tightly to more mystical or Platonic views.

The two "F"s, formations and fossils, intimately connected, were the prime focus of 17th- and 18th-century geological study.

STENO'S FOSSILS

The first major thinker to seriously examine these two major problems was a 17th-century Dane, Niels Steensen (1638–86), better known to us today as Nicolaus Steno. Born in Copenhagen, Steno was the son of a wealthy goldsmith. He trained in medicine, as so many early scientists did, and was fortunate enough to be appointed personal physician to the grand duke of Tuscany. With little to worry about, since the duke was both healthy and generous with his money, Steno had plenty of time to indulge his interests. One of those interests was fossils.

While dissecting sharks, Steno compared their teeth with so-called tongue-stones, which the ancients believed fell from the heavens at night. After convincing himself that the tongue-stones were in reality petrified sharks' teeth, Steno became an avid collector of fossils, visiting rock quarries and examining different rock formations.

His careful studies convinced Steno of the truth of the ancient argument that the fossils were the remains of living creatures preserved in rocks that had once been laid down in the sea. In 1669 he published his arguments in a book with the unwieldy title *Prodrome to a Dissertation Concerning a Solid Naturally Enclosed within a Solid.* Fossil-bearing rocks, Steno said, were created from the deposits of muddy water, both salt and fresh. Steno argued that some of the fossils, such as of sharks and sea shells, were obviously marine and so salty seas must once have covered all the inland places in which Steno found them. Others, he pointed out, represented creatures that had lived in fresh water and had probably been carried to their destinations by Noah's flood, which was generally believed to have occurred around 4,000 years before Steno's time. The big bones and teeth that were found were most likely the remains of much more recent creatures such as the elephants that had moved Hannibal's army against Rome in 218 B.C.

He also suggested a reasoned method by which gigantic layers of rocks formed from underwater deposits might collapse and create mountains and valleys. Not all the mountains were created in this way, though, according to Steno. Some were obviously volcanic, vomiting out ashes and rocks from fires burning deep within the Earth, and still others had been etched out by running waters in the same process by which cliffs and gorges were cut by running streams.

Steno also offered some general theories on the geological history of Tuscany which, he speculated, might also be applicable to the entire Earth. All in all it was an impressive, if faulty, performance, but Steno, an extremely religious Lutheran, failed to follow it up. Always a deeply troubled man, he apparently underwent some kind of dramatic spiritual crisis. Losing interest in geology altogether, he converted to Catholicism and spent his last days living in harsh and continual penance as a priest in Germany.

It is difficult to appreciate Steno's heroic intellectual effort outside the framework of his time. A few others, such as the ubiquitous English fellow of the Royal Society Robert Hooke, were attempting similar rational theories. But most late 17th-century speculations about the creation of the

Earth and its formation, some of which carried well over into the 18th century, either remained deeply religious or were awkward attempts to bind together the religious version of the story with the new scientific views. One, however, that stands out is the theory put forth by Georges Louis Leclerc, comte de Buffon (1707–88), whose extensive work in the life sciences we'll come back to in Part Two of this book.

BUFFON TESTS THE AGE OF THE EARTH

A naturalist at heart, Buffon felt his vast work *Histoire naturelle* (*Natural History*), published in 44 volumes beginning in 1749 should begin at the beginning. "The general history of the Earth," he wrote, "must precede the history of what it has produced." Most of his ideas were just a restatement of prevailing ideas of the time. But his discussion of the shaping of the Earth created a shock for the conservative minds of his French contemporaries.

"In effect," he wrote, "it appears certain that the Earth, now dry and inhabited, was formerly under the waters of the sea. These waters were above the summits of the highest mountains, since we find on these mountains and even on their peaks the remains of sea life and of shells which, compared to shells now living, are the same. We cannot doubt their perfect resemblance or identity of their species." Many of Buffon's ideas were (as we know today) incorrect. And many others were neither new nor original. But fitted together into a comprehensive (if very general) whole, they were tremendously stimulating to those 18th-century thinkers who, like Buffon, were trying to give new and rational answers to the old questions about the universe.

Borrowing an idea from the philosopher Leibniz, the Earth, Buffon said, was created when a comet collided with the molten Sun and knocked a piece from it. He ran some experiments with iron balls, which he heated, measuring the rate at which they cooled. From these results he estimated the rate at which a similar ball the size of the Earth would have taken to cool. Thus, he concluded, based on its necessary cooling time, the Earth was also very old. Perhaps, he suggested, as old as 75,000 or 100,000 years. Of course, more recent evidence points to an age closer to 4.6 billion years, but Buffon's estimate was much older—more than 10 times more—than the 6,000 years calculated by James Ussher in 1654. James Hutton, who will be discussed later in this chapter, would reach a similar conclusion about the Earth's great age a few years later (he too would greatly underestimate it, though). But Buffon was the first to reach a wide audience with the idea. Always politically astute, Buffon avoided major conflicts with the church over his ideas by offering his work only as hypotheses and managing to maintain a religious link in his chain of events.

The Earth, according to Buffon's hypothesis, had gone through a period of seven long epochs. This coincided nicely with the seven days of creation as taught in the Bible. After the initial epoch—its birth, caused by the collision of a comet with the Sun—the Earth rotated and cooled for around 3,000 years while it evolved into a spheroid. The second epoch found the Earth congealing and hardening into a solid body. This, according to Buffon's calculations, probably took around 30,000 years. During the third epoch the vapors from the gases surrounding the Earth formed a gigantic universal ocean that covered it. During this period, he explained, tidal action began to shape the Earth's development and distributed sea creatures (the origin of fossils) all over the Earth. He calculated that this process lasted nearly 25,000 years. Volcanic activity dominated the fourth epoch as the ocean began to subside over the next 10,000 years, leaving the remains of many marine creatures on high, dry land. And as land began to surface and cool, vegetation began to grow. Then, during the fifth epoch, lasting some 5,000 years, the first land animals began to appear. During the sixth epoch the land continued to evolve as vast continents began to separate from each

Buffon, who is known for his great work as a writer and popularizer in the life sciences, threw his net wide enough to include the origins of the Earth and numerous other scientific questions. In one experiment he used multiple lenses to test the claim that Archimedes had set fire to the Roman fleet by focusing the Sun's rays on the ships. Buffon succeeded in setting fire to objects at a distance of 150–250 feet, causing him to conclude that the story was feasible. (Figuier, *Vies des savants*, 1870)

other, drifting for nearly 5,000 years until they arrived at their present formation. Finally, at last, in the seventh epoch, proclaimed Buffon, humans appeared on Earth to reap the harvest of all that had gone before.

It was an audacious, neat and compact bit of speculative theory, and it offered vast appeal—especially since it was able to explain many mysteries, such as marine fossils discovered on high land and mountain tops.

Buffon tended to sidestep most scientific-religious conflicts throughout his work, though, and real progress was not made in geology until the full-scale onslaught on tradition by the Enlightenment thinkers freed scientists to follow the indications they found in the rocks and formations they observed.

As more researchers began to explore the field, it quickly became apparent to investigators that certain kinds of rocks, those termed *sedimentary*, had been laid down in parallel layers, or strata, and were formed from materials deposited underwater. The presence of fossil sea creatures embedded in those rocks helped to confirm this idea, but a disturbing problem remained. Many stratified rocks had been discovered in even the highest and most mountainous regions. Clearly, vast areas of what was now dry land must once have been under the sea, and it was during that time that the material now forming the stratified rocks was deposited. How then had the land emerged from the sea after the rocks were formed? Like many other blossoming scientific fields in the post-Newtonian era of the Enlightenment, the early days of geology were dominated by the search for mechanisms.

Two opposing theories quickly became preeminent. And, although both eventually fell out of favor, each contributed in its own way to the pioneering work of Charles Lyell (1797–1875), the persevering Scottish geologist who would lay the foundations for modern geology in the following century and give not only geologists, but biologists as well, a key to understanding the Earth and humankind's history upon it.

ABRAHAM WERNER AND NEPTUNE'S LEAVINGS

Abraham Gottlob Werner, the most famous geologist of his time, was born in Prussia in 1750. His father was an inspector in a large ironworks, and Werner spent most of his youth fulfilling his own interest in mines and minerals. After spending two years (1769–71) at the Mining Academy at Freiberg he moved on to the University of Leipzig to continue his studies. A brilliant student, he pursued his studies avidly, absorbing everything he could get his hands on about mines, minerals, rocks and mineralogy.

In 1775 he returned to the Freiberg Academy to become a lecturer. He would stay at Freiberg for nearly 40 years, becoming in his time not only the most famous but the most popular instructor at the school. A dynamic and hard-driving, "hands-on" teacher, Werner encouraged his students to travel the countryside and study rocks and mines for themselves rather than rely exclusively on the words of others. Unfortunately, he was somewhat dogmatic about his own views, and most of his students' field work was heavily indoctrinated by his ideas.

What Werner taught was a theory of the Earth that came to be known as Neptunism. The name (derived from the Roman god of the sea) reflects Werner's first basic premise: that the Earth was originally covered entirely by a vast, muddy primeval ocean. Suspended within this ocean were huge quantities of materials which, as the sea level began to decline, crystallized onto the seabed as "primary" rocks. Those rocks, explained Werner, were the universal formation that became the whole original surface of the Earth.

Abraham Gottlob Werner, professor of mineralogy at the Freiberg School of Mines, taught his theory of Neptunism, maintaining that all Earth's rocks originated from sedimentary processes. (Courtesy, Burndy Library)

Werner didn't attempt to explain where the original ocean came from, or by what mechanism it began to recede. The ocean's recession continued, though, according to Werner, and much later the first dry land (the original deposition of rocks) was exposed.

Gradually new layers of rocks, which could no longer be considered a part of the original formation, were laid down as further material crystallized from the ocean, as well as from sediments eroded away from the original Earth's surface. With the ocean retreating still further, greater areas of land became exposed, including not only these "transition" rocks, but also great mountains formed by the original (primary) rocks laid down many years before.

Then, continued Werner, massive erosion of the surface swept huge amounts of sediment back into the ocean, where it deposited and formed the strata of "secondary" rocks. Violent storms and temporary changes in the sea level caused irregularities in the depositions of the secondary rocks. Then, as the sea began to decline again, these layers of secondary rock were also exposed to submit to further erosion and deposition into the sea to become the most recent alluvial formations. Only recently, explained Werner, had the sea receded still further, allowing us to see these rocks that can only be found in the most low-lying areas.

Volcanoes, which seemed to trouble some of his contemporaries so much (some thought that all land might be volcanic in origin), were of little importance in the formation of the Earth's crust, according to Werner. He explained that they were probably caused by seams of coal burning near the Earth's surface.

Among those troubled by volcanoes (but not buying into the idea that all land was volcanic in origin) was James Hutton, Werner's contemporary and senior. Although he did not put forward his own ideas on the formation of the Earth's surface until 1788, Hutton's views, which came to be known as Plutonism, formed the second half of the 18th century's greatest geological debate.

JAMES HUTTON AND PLUTO'S BELCHINGS

Born in Edinburgh in 1726, James Hutton was a pragmatic Scotsman who, after completing an apprenticeship in law, took a degree in medicine from the University of Leiden in 1749. He never practiced, but turned to agriculture instead. Then, after a short stint as a farmer, he turned manufacturer, making a great deal of money from a factory he founded for the manufacture of ammonium chloride, after which he retired to study geology. It was not as much a round-the-barn trip as it appeared. His early

Volcanic eruptions in Europe in the 18th century focused attention of many geologists on the internal heat of the Earth and its role in the formation of the planet's crust. (Figuier: *Vies des savants*, 1870)

interest in chemistry had led him into medicine, and this interest flowed into mineralogy, as well, as he began to study the rocks and soil of his farm.

Hutton's new and, for the time, controversial theory was first published in the *Transactions* of the Royal Society of Edinburgh in 1788, although he had been working on it for more than 20 years. Hutton did not argue with Werner's observations that the sedimentary rocks had been laid down underwater, but he did dispute the claim that all the material forming those rocks had once been suspended in a vast primeval ocean that covered the entire Earth.

In Hutton's theory, based on systematic observations and thinking, he not only proposed an alternative to the Neptunists, but he also laid down an important principle, for which he is often named the founder of geology. That principle, known as Actualism, proposes that the Earth's surface was formed by forces—erosion and volcanoes—that were still active in modern times and could still be observed. It is closely linked with a principle known as Uniformitarianism, further developed in the following century by Lyell, which combines Actualism with several other factors to form many of the basic premises on which modern geology is based.

Werner had refused to speculate about the origin of the Earth and the primeval ocean (although many others, including Buffon, were doing that). His theory simply assumed the primeval ocean had existed and that conditions had once been very different from those operating today.

Hutton assumed (perhaps based on his religious beliefs) that both land and sea came into existence nearly simultaneously, proposing that we could only understand what the evidence presented to us about the Earth's "continuing" process and see it as an indefinite cycle, with "no vestige of a beginning—no prospect of an end."

The land's surface, he explained, is slowly eroded by wind, water and frost. The debris is then carried out to sea by rivers, where it is deposited on the bottom.

At the bottom of the sea, pressure and heat from deep within the Earth then "bake" the strata into sedimentary rock. Different layers of rock could be formed depending on the nature of the land erosion. Later, upward pressure on the seabed raises it through a series of earthquakes until the sedimentary rock is exposed as dry land. Much of this rock, Hutton continued, could have been deformed in the long slow process, and cracks in it might have allowed molten rock from deep within the Earth to seep upward, where, upon reaching the surface, it might form volcanoes. Igneous rock, formed by heat, might also work its way into the strata of sedimentary rock. There it would slowly cool and form crystalline rocks such as granite. Meanwhile the steady process of erosion is continuing and is already beginning to wear away the new land surface. While further erosion wears down some of the sedimentary rock and exposes the granite, new layers of sedimentary rock continue to form on the seabed. This rock, too, may then be forced upward by pressure and movements within the Earth, continuing the cyclical process, a process that continues today.

Hutton saw that this process was not only machinelike, but he recognized that changes observed in the present could account over a long period of time for observed features of the Earth's crust—the principle that became known as Actualism. Hutton proposed that these long cycles had been operating in the same slow way and at the same slow rate, throughout the Earth's history. And he saw, as well, that the Earth's history must be very long—much longer than most scientists had suspected—and that the Earth was therefore much more ancient than anyone believed.

In 1795 Hutton expanded and published his theory in a two-volume book entitled *Theory of the Earth*. The few people who found it readable (he was a terrible writer) also found it controversial. Werner's followers took it as a direct attack on Werner's Neptunian system, and conservative theologians saw it as an attack on the account of creation as related in the Bible. Still, Hutton's Plutonism (named after the god of the underworld, since much of the energy needed to drive his cyclical changes came from the internal heat of the Earth) picked up some devoted followers. A few more disciples were added in 1802 when John Playfair straightened out much of Hutton's unwieldy prose and offered Hutton's theories in a much more accessible form in his book *Illustrations of the Huttonian Theory*.

It would be nice to report that Hutton was right and his theory won the day. Neither history nor science is quite that simple, though. While Hutton's basic premise—that the geological changes in the Earth were uniform and cyclical, and operated over a long period of time—was essentially correct, his explanation of the mechanism of change was wrong. In the 19th century, Charles Lyell would pick up Hutton's Uniformitarian principles and put the pieces together to give the world a truer picture.

CUVIER AND CATASTROPHISM

The Neptunism vs. Plutonism debate dominated geology for most of the latter part of the 18th century. And matters were not helped by the intrusion of the famous French zoologist Georges Cuvier into the picture. Cuvier (1769–1832), who will be discussed further later in this book, was one of the most powerful scientists in his time. Brilliant, opportunistic and convinced of the correctness of his own opinions, Cuvier had studied fossils extensively and arrived at the conclusion that their existence could only be explained if the world had been subject to a series of major floods throughout its history. Each of these many floods, Cuvier argued, destroyed all living things on the Earth, leaving only the fossils. And after each flood, life was created anew. The view had appeal to a great many religious thinkers, especially since Cuvier explained that the last great catastrophe (his theory was later called Catastrophism) was the flood described in Genesis. During that catastrophe, Cuvier explained, God had intervened, permitting, as the Bible described in the story of Noah, some living things to survive. Obviously, Cuvier decreed, and many listened, the Earth did not undergo a continuing process of slow and gradual change as Hutton had argued, but was instead the product of a series of violent calamities.

Cuvier was so powerful that his dramatic theory of Catastrophism soon superseded the less exciting "primeval ocean" mechanics of Werner, and all but knocked Hutton and his Plutonism out of the picture. Both Werner and Hutton were specialists, not generalists, and both had attempted to treat geology as a science. Both had proposed mechanisms that they each believed fitted the facts as they observed them (although Hutton is sometimes accused of being too casual in his field work). Both were pursuing avenues that proved unproductive, even though Hutton's Actualism opened up a direction that would prove fruitful in the future. But Cuvier's prodigious reputation as a careful methodologist and seasoned theorist captured the imaginations of most scientists at the time. And, his giant shadow discouraged healthy questioning in geology for the rest of the 18th century and well into the 19th.

CHAPTER 4

WHAT HAPPENS WHEN THINGS BURN? THE DEATH OF PHLOGISTON AND THE BIRTH OF CHEMISTRY

On May 8, 1794, at the peak of the Reign of Terror in France, Antoine Lavoisier, one of the greatest scientists of all time, was hauled away to his fate at the guillotine. As one of the old, deposed royal government's tax collectors, he was seen by many as a public enemy, a member of the Fermiers Généraux (Farmers General), who made a profit by skimming a fee off the top of taxes paid to an unpopular king by a poor and downtrodden populace. On top of that offense, Lavoisier had opposed the admission of Jean-Paul Marat (a journalist who fancied himself a scientist) to the French Academy of Sciences, a move that turned out to be a disastrous tactical error because Marat never forgot the blackball. When Marat gained power in the revolutionary government, he saw a way to gain revenge and had Lavoisier arrested, tried and convicted. A key force in the Reign of Terror that washed the streets of Paris in blood, Marat was himself arrested and guillotined before Lavoisier. But the wheels of bureaucracy were not reversed, the Revolutionary Tribunal had passed its sentence, announcing, "The Republic has no use for scientists," and Lavoisier's sentence was carried out as scheduled.

It was a grim day in the history of science, for Lavoisier—with the help of four other great chemists, Scheele, Priestley, Black and Cavendish—had finally, just a few years before, begun to unravel the chaos that had reigned in chemistry since ancient times. And Lavoisier had brought the scientific revolution to chemistry just five years earlier with his *Traité élémentaire de chimie* ("Elementary Treatise on Chemistry"), published in 1789. At 51, he

Lavoisier and the Fermiers Généraux standing on trial before the Revolutionary Tribunal (Figuier: *Vies des savants,* 1870)

was at the height of his potential. In the words of Joseph Louis Lagrange, spoken on the day after Lavoisier's execution, "It took but a moment to cut off that head; perhaps a hundred years will be required to produce another like it."

COOKS OR MYSTICS

"Chemistry like all other sciences, had arisen from the reflections of ingenious men on the general facts which occur in the practice of the various arts of common life," Scottish chemist Joseph Black liked to tell his audiences. And chemistry—partly because it was so much involved with the everyday stuff of life—was, of all the scientific disciplines, the slowest to emerge from the Middle Ages, to throw off the old confused approaches and enter the post-Newtonian age. The free development of chemistry was victim, on the one hand, of the stigma of being too common and everyday (Greek philosophers, in particular, disdained any discipline that involved using one's hands—and chemistry was closely identified with cooking, the trades and medicine). It was stymied on the other by its classification as secret in the hands of alchemists who sought to transmute, or change, other substances into gold. (Some alchemists kept mum because they were naive enough to believe that they were close to a breakthrough; the rest sought silence because they didn't want to get caught in their hoax.) And chemistry, of all the sciences, resisted application of the scientific method in the hands of early experimenters. Physics had the universal laws of motion discovered by Newton; astronomy had the discovery of the rotation of the Earth and its revolution around the Sun; biology had Harvey's work on the movement of blood and the action of the heart. But chemistry had resisted any such breakthrough, understandably, given more recent 20th-century theories about the formation of compounds and the complex processes that take place at the molecular level in chemical reactions. All this was completely unknown in the 18th century; atomic theory as we know it had not even yet been conceived of.

As a result, at the beginning of the 18th century, great confusion reigned among those whose interests concerned the mixture and heating of substances. And scientists had no hope of ever making any progress until they had a handle on basic facts about air and water, the existence of "gases" and the process of combustion. It proved one of the most stubborn series of problems facing scientists in the 18th century, revolving around such basic issues that today they seem mystifying to us as questions at all—until we put ourselves in the shoes of 18th-century thinkers. The processes of combustion, calcination (the process that takes place when a substance is heated to a high temperature, but below the fusing or melting point) and respiration were recognized as crucial and interrelated, but exactly how was elusive. And many fine minds expended a great deal of time on this problem from the time of Robert Boyle and Robert Hooke throughout the 18th century.

BIRTH OF THE PHLOGISTON THEORY

Unfortunately, the first effort at establishing an integrated theory to explain combustion took science down one of those long side roads that lead virtually nowhere. The idea came, originally, from a man named Johann Joachim Becher [BEK-uhr] (1635–82), a latter-day alchemist who sold some city leaders in Holland on the claim that he could use silver to turn tons of sand into great heaps of gold. When his demonstrations didn't work, he fled to England. His ideas about matter were not much sounder. All bodies, he said, consisted of air, water and three kinds of earth, which he called *terra pinguis* (fatty earth), *terra mercurialis* (mercurial earth) and *terra lapidia* (stony earth).

In the late 17th century Georg Ernst Stahl [SHTAHL] (1660–1734) changed the name of Becher's *terra pinguis* to *phlogiston*, which he described as potent—"fire, flaming, fervid, hot"—a fluid that was released or lost by any substance when it burned, calcinated or otherwise oxidized (though that term was not yet used). In 1697 Stahl set forth his doctrine of phlogiston, and it became the framework that served to explain a constellation of puzzling chemical phenomena, the be-all and end-all of chemistry for more than 90 years.

Even though phlogiston did turn out to be a blind alley, not all consequences were negative since it stimulated endless experiments on combustion, on oxidation, on respiration and on photosynthesis.

These experiments soon revealed glaring problems, though. For one thing, a metal's oxide always weighed more, not less, than the original metal. If a substance was lost in a process, weight would normally be lost, not gained. To explain this without tossing out the theory, many chemists put forth ever more complicated explanations: Perhaps, since one could not really see phlogiston (only the flame as evidence of its exit from the burning substance), it was not a normal substance. Maybe it was more like one of the phenomena that 18th-century chemists assumed to be "subtle" or "imponderable fluids," such as magnetism, ether, heat, light and electricity, and, like them, had no weight. Maybe, even, it had negative weight, so that when it was present in the substance it actually (somehow) subtracted weight from it. And when combustion took place, phlogiston was released and the weight of the substance increased.

But, despite this nagging weight problem, most chemists of the 18th century continued to subscribe to the phlogiston theory, seeing no compelling reason to abandon it, and, nonetheless, made landmark contributions to the study of gases in particular.

Gases, chemists had begun to find, had the greatest potential for responding to scientific inquiry about chemical makeup. Breakthroughs in this area laid the groundwork for discoveries about other elements and compounds and, ultimately, in the 19th century, atomic theory. Stephen Hales, a parish priest and amateur scientist in England, made these breakthroughs possible by his invention of a device for collecting gases, the pneumatic trough. The device separated the flask in which a reaction took place from the vessel in which he collected the gas, and in this way was able to separate different gases. But he, like most chemists at the time, did not really recognize that these were essentially different substances. He just thought they were different types of air.

JOSEPH BLACK AND "FIXED AIR"

To physician-chemist Joseph Black, professor of chemistry at the University of Glasgow and, later, the University of Edinburgh, would go the honor not only of discovering carbon dioxide, but of establishing quantitative analysis.

Born in Bordeaux the son of a Scots-Irish wine merchant, Black went back to the British Isles in 1740 for his education. After completing his studies in medicine at the universities of Glasgow and Edinburgh, he returned to Glasgow to practice and teach. Highly respected by his students (of whom James Watt was one, as well as a friend), he was described by one of them as a man who "possessed that happy union of strong but disciplined imagination, powers of close undivided attention, and ample sources of reasoning, which forms original genius in scientific pursuits."

In his later years, wrote his student, "his countenance continued to preserve that pleasing expression of inward satisfaction, . . . easy, unaffected and graceful."

As a young man working on his thesis Black was investigating a major component in "Mrs. Joanna Stephens's secret remedy" for treating bladder stones (for which she sold the recipe to Prime Minister Robert Walpole for £5,000) when he discovered "fixed air." The component was calcined snails, which Black proceeded to test, using the quantitative methods he had developed. Many of the terms used in his descriptions of his experiments are unfamiliar today—quick lime, calx and so on—because his work preceded Lavoisier's highly useful reorganization of chemical nomenclature. But, in modern terms, he found that calcium carbonate converted to calcium oxide when heated strongly, giving off a gas that could then recombine with the calcium oxide to reform calcium carbonate. Because the gas could be recombined, or fixed, back into the solid it came from he called it "fixed air." Today we are more familiar with it as carbon dioxide.

Others actually had done some previous work with carbon dioxide, but Black was much more thorough. He showed that carbon dioxide could be obtained by heating, by decomposition of a mineral as well as by combustion and fermentation. He also made a breakthrough with regard to gases, demonstrating that they could be manipulated and tested in much the same way that liquids and solids could be tested. Black recognized that carbon dioxide was a natural component of the air, that people breathe out carbon dioxide and that a candle could not burn in it. Here he came upon a puzzling point. A candle placed in air in a closed container would finally go out. This was logical, since a burning candle produced carbon dioxide. But when Black used chemicals to absorb the carbon dioxide, the candle still would not burn in the closed container. He gave the problem to one of his students, Daniel Rutherford, who wrote his doctor's thesis on it.

Rutherford performed carefully controlled experiments and found that not only did a candle not burn in this gas, but a mouse could not live in it either. Like his teacher he was a subscriber to the phlogiston theory, and he believed that this air had taken on all the phlogiston it could contain (so phlogiston could not pass out of the air to the candle to burn or to the mouse for respiration). He called it "phlogisticated air"; today we call it nitrogen.

Rutherford usually receives the credit for the discovery of this noxious gas, although the details were not clarified until a few years later by Lavoisier.

Black's quantitative approach, which he passed on to his students and the scientific world at large, proved as important as his discovery of carbon dioxide. As he applied heat to the calcium carbonate in his original experiment, Black took measurements of the loss of weight involved. This gave him clues about what was taking place in the reactions—clues that chemists had not had before in their dealings with gases.

Once he had discovered fixed air, Black wrote, "a new and perhaps boundless field seemed to open before me. We know not how many different airs may be thus contained in our atmosphere, nor what may be their separate properties." In this he seemed to point forward to the coming discoveries of Cavendish and Priestley in his own time and of others, such as Baron Rayleigh and Sir William Ramsay, who would receive the Nobel Prize for their work on the constituents of the atmosphere in 1904. Black's work in the field of physics, which will be discussed in the next chapter, was equally important, with far-reaching consequences for the Industrial Revolution.

HENRY CAVENDISH AND THE COMPOSITION OF WATER

Henry Cavendish was without question the most eccentric scientist in the 18th century. Born to a famous and wealthy British family, he never had to worry about money. In fact, he once told his banker that if the banker ever bothered Cavendish about what to do with his money, the bank would lose his account, which was worth millions. Needless to say, the banker never bothered him again. But where other Cavendishes had hobnobbed with kings and participated in grand political plots, Henry's one and only interest was the pursuit of pure science.

Cavendish was a recluse to the extreme. He did not like to see people and he did not like to talk to people. (The portrait in this book, based on a painting by W. Alexander in the British Museum, is probably the only one in existence.) One day, after catching sight of a maid by chance in his house, he had an extra stairway built—one for his use alone, so that the incident would never happen again. The maid, according to one story, was fired. Unfortunately, Cavendish also published only selected bits of his work and consequently did not get credit (and the scientific community didn't get the benefit) for several achievements that he was the first to make.

Some of his work did become known to his contemporaries, though. In 1766 he presented a paper to the Royal Society on "Factitious Air," which demonstrated the existence of a distinct, flammable substance not previously

Henry Cavendish (The Bettmann Archive)

Joseph Priestley, co-discoverer of oxygen, which he called "dephlogisticated air"
(Courtesy, Burndy Library)

studied (later named *hydrogen* by Lavoisier). He also studied the properties of carbon dioxide, which he, like Black, called "fixed air."

And his most signal achievement he did publish. On January 15, 1784, he demonstrated that his Factitious Air, when burned, produced water. This was a stunning bit of news. Aristotle had maintained that water was one of the four indivisible elements of which all substances were made. But if water was formed from burning a gas, then the only explanation had to be that it was formed from the combination of two gases. It was the death knell for the Greek system of elements.

Cavendish, like Black, did extensive work in physics, which will be covered in the next chapter, but he also experimented with air, passing electric sparks through it in 1785. To explain what he did in modern terms, he forced the nitrogen in the air to combine with oxygen, then dissolved the oxide he obtained in water. (In the process he figured out how to produce nitric acid.) He kept adding more oxygen, with the expectation that eventually he would use up all the nitrogen. But a small amount of gas—just a bubble—remained behind. Now, speculating a bit, he figured that this small amount of gas must be something not encountered before, very resistant to chemical reaction, in fact, what we would call inert. Henry Cavendish had discovered argon, although it would not be verified until a century later when William Ramsay followed up.

SCHEELE AND PRIESTLEY DISCOVER OXYGEN

Joseph Priestley (1733–1804) was an English Unitarian minister and a radical in politics, which eventually created problems for him later in life. He openly supported the American colonists in 1776 in their revolt against George III. He opposed slave trade and religious bigotry. And he sympathized with the French Revolution. Most of his 150 books, one of which was publicly burned in 1785, were religious and educational. And his introduction to the study of gases may seem unusual: He began his experiments at a local brewery in the town of Leeds where he lived. He discovered that Joseph Black's fixed air produced a pleasant, bubbly drink when mixed with water—and that's how soda water was invented.

Priestley was curious and enthusiastic with his experiments, but he was not very orderly or methodical in what he chose to do. (A great believer in the role of chance in his work, he liked to say that if he had known any chemistry, he would never have made any discoveries.) But what he did do he observed very carefully. As a result, as 19th-century chemist Humphry Davy once remarked, "No single person ever discovered so many new and

SCIENTIFIC SOCIETIES

Scientific societies came in many shapes and sizes during the science-conscious 18th century. Small, informal groups like the Lunar Society, founded around 1766 by Matthew Boulton, were frequently formed among friends and colleagues (James Watt, Joseph Priestley and Erasmus Darwin were among the Lunar Society's most illustrious members). At the same time great national societies and associations also flourished, many of them still maintaining a strong influence today. The Royal Society of London (founded in 1662) and the Paris Academy of Sciences (founded in 1666) were among the most prestigious. The Berlin Academy (founded in 1700) and the St. Petersburg Academy (founded in 1724) were also highly respected. On the other side of the Atlantic the American Philosophical Society (founded in Philadelphia by Benjamin Franklin in 1743) offered the first regularly meeting organization for that recently settled region's scientific thinkers.

As is still the case today, funding was always a precarious matter. Following in the style of the Paris Academy, the academies at Berlin and St. Petersburg were composed primarily of small, elite groups of scientists, financed by the monarch and subject to the vagaries of government whim for their support. The London and Philadelphia societies were self-supportng, dependent on the private financing of participating scientists and other interested individuals.

The societies, large and small, formal and informal, formed important instruments in the spread of science during the 18th century. They not only provided an organization through which members could meet and exchange the newest scientific information, but most also published papers and journals, providing a wider distribution of scientific findings. In addition the societies became valuable in promoting international communications within the scientific community.

---

curious substances." The list includes ammonia, sulfur dioxide, carbon monoxide, hydrogen chloride, nitric oxide and hydrogen sulfide.

And one other: oxygen. In 1774 Priestley used a burning lens with a diameter of 12 inches and a focal distance of 20 inches to extract a gas from mercuric oxide. The new "air," which seemed somehow purer than normal air, caused a candle to burn brightly and vigorously. He tried the effects of breathing the new air—on mice, on plants and even on himself. He found breathing it "peculiarly light and easy." Clinging to phlogiston theory, he called this new air dephlogisticated air, the natural reverse of the name given

By the later 18th century various societies such as the Manchester Literary and Philosophical Society (founded in 1781) and the Royal Society of Edinburgh (founded in 1739 and chartered in 1783) began to fill the need for organizations that could accommodate members geographically distanced from the major world cities. In addition, after the early 1800s more specialized societies such as the Linnaean Society (founded in 1788), the Geological Society (founded in 1807), the Astronomical Society (founded in 1820) and the Chemical Society (founded in 1841) filled the need for more focused and specialized interests.

In many ways the precursor to today's scientific societies, the Royal Institution of Great Britain (founded in 1799) began to break traditions by functioning not only as a scientific society, but as a promoter of scientific events and lectures aimed at educating the average citizen in the theories and workings of science. The British Association for the Advancement of Science (founded in 1831) also furthered this endeavor by offering a spectacularly successful series of popular meetings featuring world-class scientists and attended by large audiences of middle-class interested spectators.

These associations all were intended to further the interests and traditions of scientific research, but their founders usually also had additional objectives in mind, often political, religious or, in the case of the later societies of the Industrial Revolution, economic. Though rarely pure guardians of an idealized autonomous scientific tradition, these societies nonetheless provided a priceless forum for exchange of information and encouragement of progress in the sciences.

by Black to the strange, non-life-supporting air that he had found, since its properties seemed to be exactly the reverse.

Unknown to Priestley, meanwhile, Carl Wilhelm Scheele [SHAY-le] (1742–86), a Swedish chemist, had already discovered the same gas, but news of his discovery had not been published. In fact the two deserve equal credit, since they both made the discovery independently.

In 1780, Priestley moved to Birmingham, where he antagonized the residents with his theology but found a stimulating and supportive group of scientific friends known as the Lunar Society. This group, partly organized by Erasmus Darwin (1731–1802), the grandfather of the evolutionist

Charles Darwin, met informally at dinner gatherings in each others' houses for congenial discussions of science. It became one of the most stimulating forerunners of the "think tank" in history, meeting on the night of the full moon each month so everyone could find the way home in the dark Birmingham countryside. James Watt, inventor of the steam engine and friend of Joseph Black, was there. Erasmus Darwin, always a controversial talker, often came. At the center of the group was Matthew Boulton, Watt's partner, whose optimism, commercial skill and enthusiasm for steam power provided one of the moving forces in the Industrial Revolution. The discussions usually began about 2:00 P.M. and continued until 8:00 P.M. For Priestley and the others, the lively, informed discussions of the problems of phlogiston and heat, of the composition of water, of metallurgy, electricity and astronomy served as a constant spur to work.

Priestley built an elaborate laboratory in Birmingham, reputed to be one of the best outfitted in Europe (though this may be an exaggeration), with all the latest equipment, no doubt assembled in consultation with his talented new friends. He worked quickly and wrote up his results by the fire at home, surrounded by his children—probably not the best way to concentrate, but he thought that isolating himself to write was antisocial. "My manner," he wrote, "has always been to give my whole attention to a subject till I have satisfied myself with respect to it, and then think no more about the matter. I hardly ever look into any thing that I have published, and when I do, it sometimes appears quite new to me. . . ."

Ironically, on the day of Lavoisier's untimely death, Priestley was heading for safety in America after rioting antirevolutionaries burned his house for his sympathies toward the revolutionaries in France. Already a friend of Benjamin Franklin, he also became friends with Thomas Jefferson and found an oasis from persecution in the United States. He turned down offers of positions from the Unitarian Church and a professorship of chemistry at the University of Pennsylvania, spending the last 10 years of his life quietly writing.

ANTOINE LAURENT DE LAVOISIER AND THE DEATH OF PHLOGISTON

Nineteenth-century German chemist Justus von Liebig once said that Lavoisier "discovered no new body, no new property, no natural phenomenon previously unknown. His immortal glory consists in this—he infused into the body of science a new spirit."

Known as the founder of modern chemistry, Lavoisier instilled in his colleagues a new respect for quantitative techniques, the foundation of all progress in the field. While Black and Cavendish both instituted the use of careful quantitative analysis, Lavoisier succeeded in convincing other chem-

Antoine Lavoisier, founder of modern chemistry (Figuier: *Vies des savants,* 1870)

ists of their importance. He did for chemistry what Galileo did for physics: introduced sound methodology, empiricism and a quantitative approach.

In a demonstration about how important this quantitative approach was, Joseph Louis Proust discovered in 1788 the law of definite proportions which began to hint at the theory of the atom—the idea that whole units were being exchanged in reactions. For example, a compound might contain two elements in the ratio of 4 to 1, but never in the ratio of 3.9 to 1 or 4.2 to 1, and so on. One element might combine in different proportions with other elements to produce different compounds. But each of those compounds would still obey the law of definite proportions. Carbon dioxide, for example, is made up of carbon and oxygen in proportions by weight of 3 to 8. Carbon monoxide (made up of the same elements in different proportions) is composed of carbon and oxygen in proportions by weight of 3 to 4. This quantitative discovery would prove a keystone to modern chemistry in the future.

Lavoisier was a mover in the scientific world; although his money came, to be sure, from the Fermiers Généraux, he spent it lavishly in the interest of science, and his private laboratory was a meeting place for all the major scientific figures of Europe. Thomas Jefferson and Benjamin Franklin both were warmly welcomed there. Lavoisier's wife, Marie-Anne, who married

Antoine Lavoisier and his wife, Marie-Anne, presiding at the death of phlogiston (Courtesy, Parke-Davis, Division of Warner-Lambert Company)

him when she was only 14, attended these meetings and wrote about them, illustrated them for Lavoisier's books and was always deeply involved in his work. She translated works from English for him, took notes and participated actively.

Between 1772 and 1774 Lavoisier engaged in a series of experiments and demonstrations under controlled conditions in which he tried burning various substances, including diamonds, phosphorus and sulfur, tin and lead. He burned both tin and lead in closed vessels. When these substances are heated, it had long been known, they change color and the resulting substance, known as calx, weighs more than the original metal. But when Lavoisier weighed the whole vessel in these cases, including all that was inside—air, metal, calx and the vessel itself—he found that the weight was unchanged. This meant that some part of the whole system must have lost weight, probably the air. (He discounted the idea held by some that phlogiston had negative weight.) If the air lost something, then at least a partial vacuum must exist inside the closed vessel. Sure enough, when he opened the vessel, air rushed in and when he weighed the vessel and its contents again, then it weighed more than the original. So calx must have been formed through a combination of air with metal. Hence rusting (and combustion) was a process that did not involve the loss of phlogiston; instead what occurred was a gain from the air.

The phlogiston theory was dead. The Lavoisiers organized a grand party and, in a dramatic ceremony with Marie-Anne dressed as a priestess, they burned Stahl's book on phlogiston to demonstrate the end of its grip on the science of chemistry.

And one other important result had come from Lavoisier's experiment: He also uncovered a basic principle—the law of conservation of mass, which became a "bulwark of chemistry" in the 19th century. (It was also a concept that Einstein would further refine in the 20th century).

Then in October 1774 Priestley visited Lavoisier and explained his experiments with dephlogisticated air. Lavoisier listened with interest and realized suddenly that Priestley had isolated one part of air—that the air is made up largely of two gases, one that encourages combustion and respiration and the other that does not. The phlogiston idea, as he had already concluded, was a red herring. Now the real truth seemed clear: Priestley had isolated the gas in the air that supported burning; the new gas that Priestley found was undiluted by the portion of the air that objects did not burn in. In 1779 Lavoisier announced that air is composed of two gases, the first of which—the one that supported burning—he called *oxygen* (from Greek roots meaning "to give rise to acids"; Lavoisier thought oxygen was contained in all acids, one of the rare times he was wrong). The name stuck. The other gas he called *azote*, meaning "no life" in Greek. But in 1790 it was renamed nitrogen, which is the name by which it is known today.

In a low moment, Lavoisier tried to downplay the fact that these insights had come to his attention through Priestley. He thought of Priestley as a

The lecture-demonstrations in chemistry at the Jardin du Roi in Paris—especially those given by Guillaume François Rouelle—were always exciting and presented with great flair. One of Rouelle's most famous students was Antoine Lavoisier. (Figuier: *Vies des savants*, 1870)

Lavoisier demonstrating the composition of air (Figuier: *Vies des savants*, 1870)

mere tinkerer who didn't know what he was doing: after all, he had not devoted his life to chemistry as Lavoisier had. Maybe he held some Franco-Anglo animosity against Priestley and possibly a little political antagonism, but most likely, more than anything, Lavoisier wanted to be remembered for discovering an element, which he never succeeded in doing. He did, however, interpret what Priestley had found. He played the role of theoretician and interpreter to Priestley's laboratory work, a kind of experimental-theoretical teamwork that has become more and more important in chemistry as the results of the experiments become more and more complex.

Lavoisier also clarified Cavendish's work, repeating his experiment burning the flammable gas he had found in air and forming water. Lavoisier named this gas *hydrogen*, meaning "to give rise to water" in Greek. This also fitted well with the picture he had been putting together of the new chemistry. Animals ate substances containing carbon and hydrogen, inhaled oxygen, combined them, forming carbon dioxide and water, both of which they exhaled during respiration (see Chapter 8). Here again, unfortunately, Lavoisier's character flaw reappeared and he tried to snatch the credit from Cavendish for his discovery.

The new chemistry, though, began to catch on. Although Priestley, Cavendish and Hutton never did let go of the phlogiston theory, Black aligned his thinking with Lavoisier's and so did several others.

THE LANGUAGE OF CHEMISTRY

The language we use every day is full of inconsistencies, double meanings and imprecision; that is part of the fun and beauty of language, that it can be bent and played with, made rich with analogy or funny with puns or graceful with embellishments. But for science, clarity and precision is essential. Not until chemists could eliminate imprecision in the language they used would their science truly become modern.

In 1787, when Lavoisier and his colleagues took a look at the language used in chemistry, they found a hodge-podge of confusion. The old alchemical and chemical texts drew from many languages—Greek, Hebrew, Arabic and Latin—and named substances based on a variety of analogies and impressions. Sulfur and mercury were called "father" and "mother." "Spanish green" meant copper acetate. Chemical reaction, by analogy, was called "gestation."

The new terminology was designed to reduce confusion, and the most significant aspect of its design was Lavoisier's idea that the names of the elements in a compound should be reflected in its name. So *Flowers of zinc* became *zinc oxide* (composed of zinc and oxygen) and *oil of vitriol* became *sulfuric acid*. The system also indicated relative proportions of the elements in a compound. Sulfurous acid contains less oxygen than sulfuric acid, and when combined with the oxides of metals, the resulting compounds are, respectively, sulfites and sulfates.

This transformation of chemical nomenclature played a key role in the scientific revolution in chemistry during the 18th century—and the part Lavoisier played is one of the reasons that he is considered the founder of modern chemistry.

Asked to help write an article on the history of chemistry for an encyclopedia, Lavoisier saw that the problem with discussing chemistry through the ages is that the names of substances change from country to country and from age to age. What chemistry needed was an international nomenclature (see box) that consistently reflected the composition of substances—not here a use, there a color, there a poetic fancy. And so, Lavoisier did for chemistry what Linnaeus had done for biology (as will be discussed in Chapter 6): He established a systematic nomenclature. With two other chemists, he published *Methods of Chemical Nomenclature* in 1787, establishing a clear and logical system of names that reflects composition, a system that received almost immediate acclaim (except for a few phlogiston holdouts) and is still in use today.

In 25 years, Lavoisier had established quantitative measurement as the basic tool of the chemist, killed phlogiston, established the law of conservation of mass and set up a new system of chemical nomenclature.

With Lavoisier's death in 1794, his part in the great revolution in chemistry came to a conclusion, but progress did not end there. From the foundations laid by Lavoisier, Black, Scheele, Priestley, Cavendish and, in a way, even Stahl, chemists in the 19th century were able to build an ever-more-accurate understanding of chemical elements, their nature, how they react with one another and what processes take place in those reactions. John Dalton (1766–1844) would build on Lavoisier and Black's quantitative analysis and wed it with the ancient Greek Democritus's atom to come up with the first quantitative atomic theory in 1803. In 1869 Dmitry Mendeleyev would classify the known chemical elements into the periodic table. At the end of the century, Marie and Pierre Curie would discover radiation. And the basis would be laid for theories of the electron and quantum mechanics.

Meanwhile, many of the same individuals who had done so much to bring chemistry into the modern era were also making exciting discoveries in physics.

CHAPTER 5

EXPLORING THE CHARACTERISTICS OF HEAT AND THE MYSTERIES OF ELECTRICITY

Heat had always been one of the big mysteries in physics, one that no one had come close to resolving at the end of the 17th century. The ancient Greeks had put forth three basic ideas about its nature—that it was a substance, that it was a quality and that it was an accident of common matter (a consequence of the motion of particles)—and versions of the first and last of these, especially, still competed for dominance in the 18th century. It was a difficult concept to grasp, and one of the reasons was that no one had yet found a way to measure quantities, or degrees, of heat.

So the first hurdle was to come up with a good system of measurement, providing a way to make quantitative comparisons of heat in different circumstances. In 1708 Ole Roemer [OHL-e RUHR-mer] (1644–1710), a Danish astronomer, was the first to recognize that a thermometer needed two fixed points, setting the point at which snow melts as one and the temperature at which water boils as the other. The Dutchman Daniel Fahrenheit used Roemer's scale, with some adaptations, for thermometers he designed in 1714, substituting mercury for alcohol. This meant that temperatures above water's boiling point could be measured because mercury has a much higher boiling point than alcohol. The Swedish astronomer Anders Celsius, meanwhile, used the same fixed points, separating them by 100 units for the centigrade scale he devised in 1742. And a fellow countryman, biologist Carolus Linnaeus, inverted his thermometer, setting the boiling point at 100 and the freezing point at zero, the arrangement still used today on the Celsius thermometer (also called centigrade), used by scientists all over the world.

HEAT AS A FLUID

In the 18th century the favored theory of heat was established by Hermann Boerhaave (1668–1738), who thought of it as a special sort of matter. This doctrine fitted in well with the theory of phlogiston (see Chapter 4), which, like light and electricity, was thought to be an "imponderable," or weightless, fluid. Though Lavoisier demolished the theory of phlogiston, he continued to think of heat as a fluid that could be poured from one substance to another, and he named it *caloric*, including it in his *Traité élémentaire de chimie*, published in 1789. It was a theory that served the 18th century well, and by the end of the century Pierre Simon Laplace (see Chapter 2) had incorporated the concept of caloric in a new, complex general picture of matter. His mathematical analysis greatly contributed to the prestige of the theory.

THE ESTIMABLE DR. BLACK AND HIS FRIEND JAMES WATT

In the 1760s Joseph Black, the physician and chemistry professor in Glasgow and Edinburgh, also became interested in the nature of heat. The question loomed particularly large in industrial Glasgow and Edinburgh, where the union of Scotland and England in 1707 had produced a healthy economy with good markets for the local whiskey industry. The big distilleries used large amounts of fuel and generated huge amounts of heat to turn liquids into vapors; then those huge amounts of heat had to be removed to condense vapors back into liquids. To manage the distilleries economically, knowing the exact amount of heat involved in these processes was an absolute necessity. In fact the need to remove great quantities of heat from the vapor had a direct effect on the profitability of a distillery.

Black often said he couldn't understand why the distillery managers had never paid much attention to the scientific principles involved, when those principles clearly had such an important impact on their livelihood. But this relationship between pure science and technological advancement is traditionally too little recognized and much too little supported. Even today research and development departments are nearly always the first to be cut back when a business is trimming its costs. And universities are always a target in lean times.

Black didn't publish his lectures, in which he discussed his ideas most thoroughly, but his editor, John Robison, published material from Black's notebooks and his own observations:

Joseph Black, who made important contributions regarding latent heat, specific heat and the distinction between heat and temperature, in addition to his work on "fixed air" (carbon dioxide) (Courtesy, Burndy Library)

> *Since a fine winter day of sunshine did not at once clear the hills of snow, nor a frosty night suddenly cover the ponds with a thick cake of ice, Dr. Black was already convinced that much heat was absorbed and fixed in the water which slowly trickled from the wreaths of snow; and on the other hand, that much heat emerged from it while it was slowly changing into ice. For, during a thaw, a thermometer will always sink when removed from the air into melting snow; and during severe frost it will rise when plunged into freezing water. Therefore, in the first case, the snow is receiving heat, and in the last, the water is allowing it to emerge again.*

In 1762, at a meeting of the University Philosophical Club, a group of professors that met informally in Glasgow, Black discussed his insights further. Ice does not change temperature when it melts, he pointed out, but material in the vicinity of the ice gets cooler, even though the temperature of the ice does not get warmer. What happens to the heat? Does it disappear? Fahrenheit had once observed that water could be cooled below the temperature of melting snow without freezing, although any disturbance of water cooled in this way caused it to immediately freeze rock solid. When this happened, the temperature actually rose! So, as the water froze, that is, as it changed state from liquid to solid, it gave off heat. Water, Black saw, remains liquid because it contains a certain amount of heat; when the heat is removed, the fluidity disappears and instead of liquid water you have solid ice. Because the heat present in the fluid water did not register on a

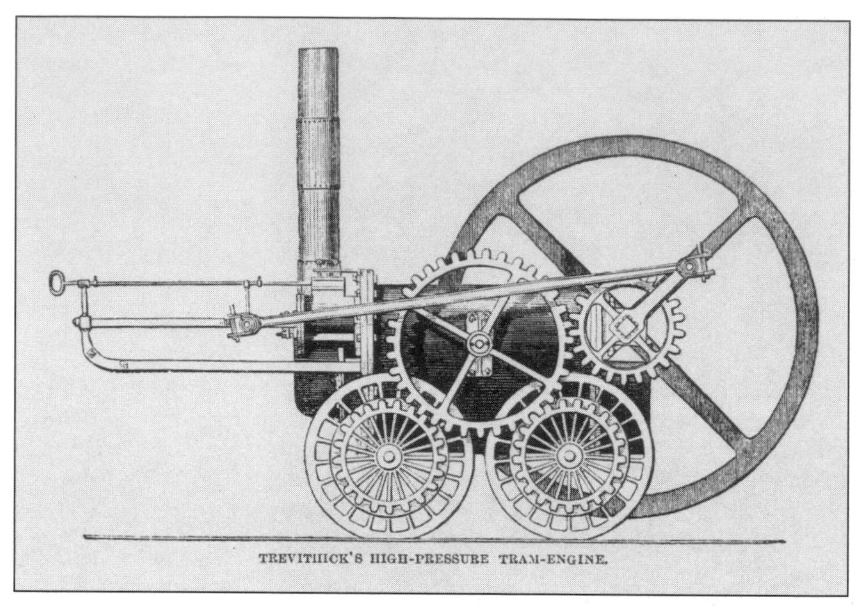

A high-pressure tram engine, designed by Richard Trevithick, probably similar to the steam carriage he built in 1801, based on James Watt's steam engine (Samuel Smiles, *The Life of George Stephenson and of His Son Robert Stephenson*, 1868)

thermometer, Black called it "latent" heat, to indicate that it was present but not measurable by the usual means.

Black also figured out a way to measure this latent heat. He measured the amount of heat required to melt a certain quantity of ice. Then he applied that amount of heat to the water produced by the melted ice and found that its temperature rose 140 degrees F.

Between 1762 and 1764, Black worked on the natural extension of the concept of latent heat in ice—the idea that something similar must happen when water is converted to steam. He found that it took about five times as long, with fire at the same strength, to transform boiling water to steam as it did to bring water to the boiling point.

At this point some extra help and a new friend entered the picture: James Watt, an instrument maker for the university. Watt devised apparatus for demonstrations and experimental proofs of the latent heat concept for Black's classes. And based on the theoretical insights he gained from Black, Watt succeeded in inventing a new apparatus for a steam engine he was repairing: a separate condenser. This single invention turned out to be the key to making steam engines efficient enough to become economical sources of energy for both transportation and industry. Fueled by coal or coke, James Watt's practical steam engine freed factories to be located

COOK, THE NEWCOMEN ENGINE AND THE INDUSTRIAL REVOLUTION

One of the most important discoveries of the early 18th century was that coke, made from coal, could be substituted for charcoal, made out of wood, to power blast furnaces. The blast furnaces produced pig and cast iron for manufacturing but, prior to the discovery that coke could be used for fuel, each furnace could consume as much as 200 acres of valuable wood for each year's operation. The discovery, generally attributed to the Quaker iron master Abraham Darby in 1709, had its first practical application in production at his furnaces in Coalbrookdale, Shropshire, England, where Darby manufactured iron boilers for the Newcomen engine invented in 1705. The engine, invented by the English blacksmith Thomas Newcomen, helped to pave the way for the Industrial Revolution. The simple device, using a jet of cold water to condense steam entering a cylinder, created atmospheric pressure to drive pistons. By 1712 the engine was in wide use to pump water out of coal mines, allowing for the greater production of coal from flooded colliery galleries.

anywhere, away from the riverside locations they had needed for water power. The steam engine soon became a cheap source of power for nearly every industry, from coal mining to ironworks to textile plants (and in the following century, railways and steamships).

Black was enormously pleased and loved to tell his students about Watt's accomplishments. When Watt took out a patent in 1769, he was on his way to well-deserved rewards. "Dr. Black," wrote Robison, "would scarcely have been more gratified, had those advantages accrued to himself. . . . both the friends considered that period of successful investigation as among the most fertile of enjoyment of any part of their lives."

Earlier, Black had also showed that equal masses of different substances require different ("specific") quantities of heat to bring them to the same temperature. Or, put another way, when equal weights of two different substances are brought to equilibrium from two different temperatures, the equilibrium temperature is not necessarily at the midpoint. That is, the same quantity of heat applied to the two different substances produces a characteristic temperature change for each substance. Black was a subscriber to the fluid theory of heat, and as he began to work out what he called "specific" heats for various substances, he became more convinced than ever that the dynamical or kinetic theory of heat proposed by Bacon (and later reframed by Rumford) was inconsistent with the existence of specific heats. Unfortunately, this opinion turned out to be an example of how sound science can sometimes seem to fail to corroborate a valid theory.

COUNT RUMFORD AND HEAT AS MOTION

The man called Count Rumford began life as Benjamin Thompson, born in 1753 in Woburn, Massachusetts. It was the beginning of a strange and eccentric life, not always honorable, but definitely interesting. As a youngster he nearly killed himself making explosives for a celebration of the repeal of the Stamp Act while working at a local retail shop in Salem. He then worked awhile at another retail store in Boston after his recovery, and at 19 he married a rich widow. He moved with her to an estate in Rumford (now Concord), New Hampshire, but the situation became complicated when the Revolutionary War broke out and Thompson began spying on his neighbors for the British. Presumably things were getting a little uncomfortable around home, so he hopped a British military ship heading out of Boston, leaving his wife and child behind.

While in England, Thompson worked for the British secretary of state for the colonies (where his knowledge of America was much appreciated) and became acquainted with Sir Joseph Banks, then president of the Royal Society. Through Banks, Thompson met the leading scientists of the day.

He returned to the colonies before the end of the war, however, as a lieutenant colonel in the king's forces. Unfortunately for him, the British lost. He was forced into permanent exile, leaving immediately for England, where his opportunistic character emerged again as he took bribes and possibly even spied against the British for the French.

In 1783 he received permission from George III to head for the Continent, where he went to work for Elector Karl Theodor of Bavaria. There he served well in various administrative positions, including minister of war and state councillor. He set up workhouses for the homeless, introduced Watt's steam engine and the potato to the Continent and accomplished other positive goals. The elector was pleased and Thompson became a count of the Holy Roman Empire in 1791, choosing to use the name Rumford, probably out of nostalgia for the estate he had left behind in New Hampshire.

While in Bavaria, Rumford, who had always been interested in science, became deeply involved with the nature of heat. While boring a cannon in Munich in 1798, he observed that the metal got very hot as the boring machine chiseled the hole—so hot that it had to be cooled with water. Caloric-theory backers would explain this by saying that the weightless caloric fluid was being released from the metal as the shavings were scraped away. But Rumford noticed that the heat never let up as long as the boring continued, and if one measured the amount of caloric released during the boring, the total released would be enough to melt the metal if (somehow) it were poured back into the brass. He also noticed that if he used instruments so dull that no shavings were cut from the metal, the metal got even hotter—not less so, as would be expected if it was the process of breaking up the metal that caused the release of caloric.

So Rumford suggested a kinetic theory of heat, namely, that the mechanical motion of the boring tool was converted into heat. Heat, therefore, he said, was a form of motion. It was an idea hinted at before him by Francis Bacon, Robert Boyle and Robert Hooke.

Rumford's idea would soon gain some strong adherents, who rejected the caloric theory in favor of the kinetic. But, as the century closed, caloric theory remained the favorite of most physicists and chemists—no doubt still influenced by the powerful opinion of Antoine Lavoisier and by the mathematical support for caloric. But Rumford, it turned out, was right, though it would not be shown until James Joule (1818–89) was able to provide a mathematical quantity for how much heat was produced by a given quantity of mechanical energy.

Rumford returned to England in 1799, the negative aspects of his character finally having worn thin in Bavaria. In England he was admitted to the Royal Society for his achievements and founded the Royal Institution that same year. Thomas Young and Humphry Davy, two rising young physicists, became lecturers there, and Davy, in particular, was enthusiastic about Rumford's kinetic theory, publishing results of an experiment he had done that he thought supported it. But most physicists remained unconvinced until James Maxwell finally established kinetic theory in 1871.

In 1804, Rumford (whose first wife had died) married Lavoisier's widow, Marie-Anne, who was wealthy and, of course, famous (she kept Lavoisier's name after her marriage to Rumford). He was 50, she about 47. But the marriage did not go well; they began to quarrel immediately and they lasted together only four years, with Rumford muttering that Lavoisier was lucky to have been guillotined. Rumford, of course, as one historian puts it, was no daisy, but he did come out smelling like a rose—with half of Marie-Anne Lavoisier's fortune.

His American daughter by his first wife joined him in 1811 and looked after him in his later years. And, though he was a crusty and often unscrupulous character, in addition to the contributions his work made to science, he also contributed numerous practical inventions—including a double boiler, a drip coffee pot and a cooking range—that he purposely left unpatented (free to be used). He also left most of his estate to the United States when he died in 1814, as well as an endowment for a professorship in applied science at Harvard University.

ELECTRICITY: THE GREAT PARLOR GAME

People have known a little bit about electricity for a long time. A piece of amber, both the ancient Greeks and the ancient Chinese had found, would attract light-weight materials, such as a feather or a scrap of cloth, if one rubbed it. The Greeks called this fossil resin substance *elektron*. But no one knew a lot about it.

In 1600, in his book on magnetism, William Gilbert distinguished between magnetism, the power of a lodestone to attract iron, and what he called *vis electrica*, the ability of amber (and other substances he found, such as jet and sulfur) to attract objects when rubbed. He was the first to suggest that this property was not intrinsic to the amber, jet and sulfur but was a fluid, either produced or transferred by the rubbing. But he didn't discuss the *vis electrica* much because he thought it was trivial.

And trivial it remained for a long time. In the 17th century the Irish scientist Robert Boyle enjoyed a pleasant diversion as much as anyone, and he recounted noticing

> *. . . that false locks of hair, brought to a certain degree of dryness, will be attracted by the flesh of some persons. I had proof in two beautiful ladies who wore them; for at some times, I observed that they could not keep them from flying to their cheeks, and from sticking there, though neither of them had occasion for or did use paints.*

And Boyle's contemporary, Otto von Guericke of Magdeburg in Saxony, carried out electrical experiments using a rotating sphere of sulfur as an electrical machine. Created by fusing sulfur and other minerals inside a glass globe used as a mold and later removed, this machine was "as large as an infant's head." To provide an axis on which to spin it, von Guericke bored a hole through the center through which he inserted an iron rod with a handle attached to turn it. Holding one hand against the globe, he could crank it with the other, and the friction electrified the sphere, which would then attract other objects. von Guericke found that he could transfer this electricity to other objects, such as another ball of sulfur. And he also noticed another interesting point: the objects that were initially attracted to the sulfur globe were repelled by it once they touched it.

By the 18th century the use of electrified glass globes and rods became a party diversion all over Europe. Guests amused themselves in shocking each other; moving light objects, such as feathers; and making each others' hair stand on end.

Naturally scientists wondered what caused this phenomenon. They suspected that it probably was another "imponderable" fluid—as many of them also classified heat and phlogiston, the substance thought to cause combustion. To explain the attraction/repulsion that von Guericke had noticed, the most popular theory was a two-fluid theory. One fluid repelled; the other attracted. Rubbing a glass rod or globe with fur transferred some of the fluids, creating the electrical charge. Opposite fluids then attracted each other (in the same way opposite poles attract in magnets).

One of the first breakthroughs came in 1729, when Stephen Gray found that when he electrified a long glass tube with corks at either end, not only the tube but also the corks became electrified. He had discovered electrical conduction. And a strange contraption called the Leiden Jar, invented in 1746 in Leiden in the Netherlands by Pieter van Musschenbroek, led to even more fruitful experimentation. It was a glass jar coated with metal both inside and out, and it was basically a storage container (named a "capacitor" a half-century later by Alessandro Volta) that could store large quantities of static electrical charges produced through friction. If one wanted to discharge (cause the electricity to be released from) a charged Leiden Jar, all one had to do was bring one's hand near the center rod, and in the early days of electrical research, this brought many a researcher a nasty jolt. A piece of metal brought near the jar would cause tiny jagged sparks to jump across the gap, giving off sharp, crackling noises.

Like the glass globes and the glass rods, the Leiden Jar became a great conversation piece at social gatherings. But its invention also marked the beginning of serious research on the nature and characteristics of electricity.

BENJAMIN FRANKLIN, ELECTRICAL DIPLOMAT

American scientist Benjamin Franklin (1706–90) was known throughout the world by the time of his death as a national leader in his country, an international diplomat, an ingenious inventor and a weaver of homespun wisdom. He was also world-renowned for his fertile and stimulating mind—he was friends with many European scientists, including Priestley and Lavoisier—and, especially, his work with electricity, some of it exceedingly risky.

When he retired from his business a wealthy man at 42, Franklin was free to become even more involved in the electrical studies he had begun as early as 1746. He developed a theory that frictional electricity involved a transfer of "electric fluid," leaving surfaces "positive" or "negative." Positive charge was most likely an extra amount of fluid, and negative charge would be a shortage of it. Although the fluid theory didn't last long beyond the 18th century, the idea of positive and negative charges did. This "one-fluid theory" broke with the commonly accepted "two-fluid theory."

THE ENTERPRISING MR. FRANKLIN

In addition to his active scientific and political pursuits, Benjamin Franklin was also America's first major publisher. Beginning in 1732 his *Poor Richard's Almanack* continued through 25 editions over the next 25 years, becoming in its time the most popular publication (after the Bible) in the colonies. An agricultural handbook, the *Almanack* was chock-full of recipes, casual and useful information on such subjects as health and personal hygiene, and down-home folksy sayings usually attributed to "Richard Saunders," in reality Franklin himself.

Reorganizing the rough-and-tumble colonial postal service after becoming postmaster general, Franklin improved the circulation of his *Almanack* as well as mail service in general by doubling and tripling postal service in rural areas, increasing the speed of delivery and improving roads from Maine to Georgia into a system that became known as the King's Highway.

His success in business, including such practical inventions as the Franklin stove, allowed him to retire in 1748, intending to devote the remainder of his life to his scientific pursuits. These included his famous experiments with electricity, as well as ideas about light that rejected the particle theory and foreshadowed the work of Thomas Young in the early 19th century.

After founding the American Philosophical Society in 1743, he also founded, in 1749, a college that would later become the University of Pennsylvania. He traveled to Britain as the agent of the Pennsylvania Assembly in 1757 and crisscrossed the Atlantic numerous times, attending meetings of the Royal Society, while at the same time campaigning for the independence of the American colonies as their foremost spokesman in Britain.

Franklin is perhaps best known as one of the drafters of the American Declaration of Independence in 1776 and one of its most famous signers. He also succeeded in securing French aid for America during a diplomatic mission to France (1777–85) and was also instrumental in the final negotiations for peace at the end of the Revolutionary War, which guaranteed America's independence in 1783.

The United States of America's first world-class figure, Franklin gained such great popularity after the war that he soon found himself once again drawn into civic and political activities. He continued to serve in various capacities, including his important involvement in the ratification of the United States Constitution, until his death in 1790.

Benjamin Franklin, America's first world-class figure, was an accomplished scientist and inventor, as well as statesman. Bifocals, which he invented in 1784, numbered among his many triumphs. (The Bettmann Archive)

Franklin also recognized the law of conservation of charge, which says that for every negative charge created, there must be an equal amount of positive charge. And all negative and positive charges in the universe must balance perfectly. So if one rubs a balloon against a wool sweater, the balloon picks up a negative charge—but it also leaves a positive charge on the sweater. Then when one places the balloon on the wall, it will stay in place because its negative charge attracts the slight positive charge already present in the wall. Along with Franklin's one-fluid theory, the law of conservation of charge helped explain what made the recently invented Leiden Jar work.

The Leiden Jar's capacity for accumulating greater charges made possible all kinds of experiments, uses and stunts—and Franklin was delighted with it ("that wonderful bottle . . . that miraculous jar!" he once remarked). So much so that in 1749 Franklin and his friends decided to throw a party on the banks of the Schuylkill River. The party's theme was electricity, its uses and wonders. They planned to send a spark across the river through the water, kill a turkey by electric shock (which tenderized the meat) and roast it on a fire kindled by the "Electrified Bottle." The day ended, however, on a shocking note, as Franklin recounted in a letter to his brother John:

> *Being about to kill a turkey from the shock of two large glass jars, containing as much electrical fire as forty common phials, I inadvertently took the whole through my own arms and body, by receiving the fire from the united top wires with one hand, while the other held a chain connected with the outsides of both jars. The company present (whose talking to me, and to one another, I suppose occasioned my inattention to what I was about) say that the flash was very great and the crack as loud as a pistol: yet, my senses being instantly gone, I neither saw the one nor heard the other; nor did I feel the stroke on my hands. . . . I then felt what I know not well how to describe, a universal blow throughout my whole body from head to foot, which seemed within as well as without; after which the first thing I took notice of was a violent quick shaking of my body, which gradually remitting, my senses as gradually returned.*

The crack and crackle and the jagged shape of the sparks set Franklin to pondering the relationship between the static electricity in the Leiden Jar and the lightning in the sky, which led him to his most famous (and dangerous) experiment. In a thunderstorm in 1752 he flew a kite that had been especially prepared with a pointed wire attached to a silk thread (silk conducts an electrical charge very well). The idea was that the silk thread would conduct the electricity from the sky (assuming there was electricity in the sky) to the ground, where Franklin had attached a metal key. He watched the sky for the right moment, then, as he saw lightning flickering in the clouds, he reached his hand up to the key. A spark leaped across the gap just as the spark from the Leiden Jar did. Franklin also was able to charge the Leiden Jar from the lightning. He had proved that lightning was electrical in nature, and he was elected to the Royal Society.

Franklin was, however, extraordinarily lucky. The next two men who tried to duplicate his experiment were killed.

Always practical, Franklin put his experience to immediate use, inventing the first lightning rod, and by 1782, 400 houses in Philadelphia, where he lived, were protected by lightning rods. He also rigged bells to ring in his house whenever charged clouds passed overhead, so that he could take the opportunity to collect electrical charges or otherwise experiment.

COULOMB'S LAW

Charles Augustin de Coulomb (1736–1806) established another major electrical law in 1789. He decided to measure electrical force using two electrically charged cork balls attached to a rod that, in turn, he suspended from a wire. Nearby he placed two other cork balls having an opposite charge. He knew exactly how much charge each ball had, and he could calculate the force of attraction between the balls by the amount of twist in the wire. The results amazed him and everyone else. He found that the electrical force between two charges depends on the strength of the two charges. That is, the bigger the difference between the two electrical charges, the stronger the attraction between them. Also, he found, the farther apart they are, the less the attraction. If they were twice as far apart, the attraction would be only a fourth as much; at a distance three times as far the amount of attraction would drop to one-ninth. And so on. These observations he summed up in Coulomb's law, stating that the force between two electric charges is proportional to the product and inversely proportional to the square of the distance between the charges.

Coulomb and his colleagues were stunned to realize that this inverse square relationship exactly paralleled Newton's law of universal gravitation. Based on Coulomb's work it was clear that gravitation and electricity worked very similarly. While he was at it, Coulomb made a similar study of magnetism and found that magnetic force also follows an inverse square law—very exciting news, for it demonstrated that all three of these fundamental forces follow similar laws. The universe, indeed, must operate by a simple, neat and orderly set of principles. As the 18th century drew to a close physicists must have felt a sense of excitement and anticipation about discoveries to come, especially in electricity, that "trivial" subject that all of a sudden seemed to be anything but.

The twilight of the 18th century glowed with the dawning of a coming era, a time when electricity would be harnessed by humans and set to work. By 1800 Alessandro Volta, an Italian professor of physics, would invent the first electric battery, which would enable scientists to make electricity and study it more effectively under laboratory conditions. And by 1831 Michael Faraday, an English physicist and chemist, would produce electricity by moving a magnet through a coil of copper wire, producing the first electric generator dynamo.

THE EIGHTEENTH-CENTURY LEGACY TO THE PHYSICAL SCIENCES

In the physical sciences, the legacy left by the 18th century to those who followed included several quiet but nonetheless key contributions: the new recognition of the importance of quantitative analysis; the continued practice of Galileo's principles of rigorous methodology, observation and experiment; and the development of experimental skills.

In addition, theoretical and experimental advances were made by the 18th-century philosophes and experimentalists in every area of the physical sciences.

In astronomy, the discovery of a new planet, the measurement of the distance to the Sun, and the establishment of new ideas about nebulas and galaxies all revealed that both the Solar System and the universe were much larger and more complex than anyone had previously imagined.

Geologists, struggling against ideological conflicts with biblical accounts of the creation of the Earth, nonetheless made extensive studies of rock strata and other geological formations, forming a wealth of theories about the Earth's history. At the end of the century, they stood temporarily stymied by the widespread acceptance of Cuvier's theory of catastrophism, but James Hutton had already put forth the beginnings of a doctrine, uniformitarianism, that would prove more productive.

The 18th century had seen a veritable scientific revolution in the field of chemistry, completely transforming that area of inquiry from its medieval roots. Black and Lavoisier had established new quantitative methods; Priestley, Scheele, Cavendish, Lavoisier and others had all contributed to a transformed understanding of gases; and a new scientific nomenclature brought a much-needed precision to the language of chemistry.

In physics, Newton's law of universal gravitation was validated by successful measurement of the shape of the Earth. And, despite an ongoing struggle to come to grips with the nature of heat, with progress hung up on the lingering theory of heat as a fluid or "caloric," discoveries about the characteristics of heat and methods for measuring it made possible some of the technological breakthroughs of the Industrial Revolution. The 18th century also saw the first scientific studies of the phenomenon of electricity, including the invention of the first capacitor (the Leiden Jar), and exploration of some of its characteristics.

Much, though, remained to be done in the physical sciences. What was the true nature of electricity? What was magneticism? What was heat? What was light? What mystery lay at the heart of chemical reactions? How far beyond our dreams did the universe really extend? Was the Earth itself young or old, and what forces had really shaped it?

The stage was set for the century to come. The wonderfully complex mystery story had many more acts. Perhaps in fact it never ends. For now, though, the scientific investigators of the 18th century had, through their interrogations of nature, opened up a provocative new list of questions in the physical sciences. It was the turn of the next generation of investigators now to try to find new answers. That, though, is another story.

In the life sciences, meanwhile, the spirit of the Age of Reason and the Newtonian Revolution were providing new structure to complex lines of investigation.

PART TWO

THE LIFE SCIENCES IN THE EIGHTEENTH CENTURY

CHAPTER 6

LINNAEUS: THE GREAT NAME GIVER

Plucked from the early days of the 16th century and dropped into the 18th, a bewildered time traveler would probably no longer recognize the world. The combined brilliance of the great 17th-century thinkers had changed the face of the physical universe—at least as people saw it. The Earth was no longer the center of the universe, but just another planet in orbit around the Sun. The Sun itself, as well as the entire Solar System, was in movement. Gone were the beautiful and perfect spheres of Aristotle and Ptolemy. The stars moved. The sky was filled with what appeared to be other galaxies, and all those, said the astronomers, were in motion. Within those galaxies, some speculated, other solar systems spun around their suns. Everything, it seemed, was movement—unceasing, mathematically predictable movement.

Only in the field of natural history—what today we call the life sciences—would the bewildered time traveler find familiar ideas. By now the physical scientists had accepted motion and not fixity, or permanence, as their guiding principle. But for those who studied living forms—the natural scientists—one ancient idea still reigned supreme at the beginning of the 18th century: the Great Chain of Being. This was the premier concept dominating early 18th-century biology: the idea that living forms, created by God, were fixed and forever the same, each and every form of life part of a grand design, like a rung on a ladder or a link in a chain.

Originating with the ancient Greeks and developed further by Christian philosophers, this idea of a Great Chain of Being held that everything that could exist already did exist. And had always existed. That everything, from inanimate objects to living things, whether insect or fish or human, formed an unchangeable and permanent link along a chain that was arranged in a hierarchy from the lowest to the highest. Plants were higher along this chain than rocks. Animals were higher than plants, certain animals were higher than other animals, and humans were higher still. For many the chain stretched even farther—beyond man to the hierarchy of angels, extending in ever-increasing perfection toward God. Each link on the chain, whether insect, chimpanzee, or human, was permanently fixed in position, and each link had an inferior on one side and a superior on the other. Each creature's position on the great chain was created by God at the beginning and could never be altered or changed. Furthermore, according to believers in this view, no new species had been created after the creation, and none had become extinct. For either to have happened would have altered the links in the chain—an impossibility, since God had created it perfectly in the beginning.

Carl von Linné, better known by his Latinized name, Carolus Linnaeus, was a devout believer in the great chain. But, as so often happens to scientists, his greatest work furthered human knowledge while working against his own beliefs. For the days of the concept of the Great Chain of Being, with its insistence on permanence and immutability, were numbered, even in the 18th century, and the major work of Linnaeus—a massive, non-hierarchical system of classification—contributed in some ways to its demise.

CAROLUS LINNAEUS, TRUANT STUDENT

Born in southern Sweden on May 23, 1707, Linnaeus was the son of a poor clergyman who named his family after a gigantic linden tree that grew near his house. While Linnaeus's father hoped that his son would follow in his footsteps and prepare for the ministry, Linnaeus was not much of a student and didn't show much interest in theology. He preferred to spend his days skipping classes. Not that he wasn't curious; it was just that his curiosity centered on areas not necessarily taught in school: He had fallen in love with botany.

The situation caused a good deal of stress and argument for Linnaeus at home, and as a result most of the boy's childhood was dismal. But fortunately he found a friend in the rector of the school who perceived the youngster's brightness and loaned him books on botany. A physics teacher also came to the rescue by diplomatically pointing out to Linnaeus's father that his son was bright and quick to learn when the subject interested him. And, the teacher suggested, his interest in botany might also be channeled into a career in medicine. His intervention seems to have led to a family compro-

Carolus Linnaeus (1707–78), founder of biological taxonomy (Figuier: *Vies des savants,* 1870)

mise, and he even took young Linnaeus into his home and helped prepare him for entrance into the university at nearby Lund.

At Lund, however, Linnaeus did not fare well. He really wasn't interested in medicine. And if he had been, he would have been deeply disappointed. The school was in financial trouble and had only one medical instructor on the faculty. Boredom quickly set in and once more Linnaeus began cutting his classes and taking to the fields.

Fortunately, though, Linnaeus had another lucky break. He had secured boarding with a member of the faculty, Kilian Stobaeus. One night Stobaeus caught Linnaeus sneaking into the library to read the professor's botany books. When Stobaeus asked for an explanation, Linnaeus tried to assure his host that he had not intended any harm. In the process, he revealed such an extensive knowledge of botany that Stobaeus gave him complete run of the library as well as his encouragement and friendship.

Despite Stobaeus's friendship, though, Linnaeus was unhappy at Lund. So, bad grades and all, he drifted to the University at Uppsala, 40 miles north of Stockholm, where he was drawn by Uppsala's famous botanical gardens. But he was again disappointed. Once a proud school, Uppsala, like Lund, was also suffering through lean financial times. The botanical gardens were in terrible condition and the medical facilities were so impoverished that the school no longer even had a laboratory of its own but was reduced to holding lectures in chemistry at an apothecary shop.

By then so poor and ill-fed that he was suffering from scurvy, painful headaches and malnutrition, Linnaeus might have dropped out from Uppsala except for another fortunate meeting with an older teacher. As he was cutting his classes one day (a habit he had again fallen into), studying what was left of the flowers in the botanical gardens, Linnaeus bumped into a theology professor, Olaf Celsius, who had a keen avocational interest in botany. In a fortunate replay of his previous experience at Lund, instead of being lectured or reprimanded for cutting classes, Linnaeus was offered the use of Celsius's extensive botanical library.

Linnaeus's meeting with Celsius resulted in an invitation to live with the older man. He also began to make a little money tutoring other students in botany, and he was slowly starting to build a reputation among the Uppsala faculty for his botanical knowledge.

After reading a review of a paper about the sexuality of plants, Linnaeus began some studies of his own, particularly on plant stamens and pistils. The result was a short paper that he presented to Celsius as a New Year's Day present. Celsius was so impressed that he had copies made and circulated. When one of the copies was read at the Swedish Royal Academy of Science, the academy recognized its worth and ordered it to be officially printed under the academy stamp. It was Linnaeus's first taste of true accomplishment.

From that point on, Linnaeus's attitude and fortune changed dramatically. He immediately applied for and received an appointment as a lecturer on the plants in the botanical gardens.

His career was beginning to look brighter, but his sudden change in fortune may have begun to go to his head. Always individualistic, he became even more eccentric and self-centered; always pious (despite not wanting a career in the ministry), he became self-righteous and moralistic. Now, too, his ambition was burning. After years of trying to justify himself and his choices to others, he was at last accepted as a legitimate botanist.

The taste of success was sweet. So sweet in fact that he decided to secure it further with some dramatic achievement that would reach beyond poor, impoverished Uppsala and raise him head and shoulders above his competitors, even those with better academic credentials.

JOURNEY TO LAPLAND

In 1732, aided by a small grant from the Swedish government, Linnaeus set off on a field expedition to study plant life in the rugged terrain of Lapland, the far northern regions of Scandinavia, above the Arctic Circle. The experience was the most dramatic of his life, and one he referred to often whenever he wrote about himself (which was also quite often: he wrote four separate autobiographies, each different from the preceding one). For five months, poorly equipped (the entire expedition had been financed for $100) and struggling against the hostile elements, Linnaeus tramped the wild northern terrain, collecting notes and plants wherever he went. Traveling

Linnaeus tumbling into the depths of a crevasse in Lapland (Figuier: *Vies des savants*, 1870)

largely on foot, he slept wrapped in animal skins, which he also used as an overcoat to protect himself from the cold. He trekked up and down the river basins along the Gulf of Bothnia, working his way slowly north, wading through frigid, knee-deep water in the icy marshlands, hiking through forests and scaling mountainsides.

All in all, Linnaeus covered 4,600 square miles in northern Scandinavia, mostly on foot in the cold lands of the north, and he returned with an impressive collection of more than 100 new specimens. He also brought back a colorful Lapp costume, which he later took to wearing on public or official occasions to remind everyone of his adventures.

Returning to Uppsala, his reputation enhanced but his pocketbook still meager, he continued his lectures to ever-increasing and appreciative audiences. He was also fortunate enough to fall in love with the daughter of a wealthy doctor, who not only heartily approved of the adventurous and pious botanist but gave him enough money to send him to a small Dutch university to finish his medical degree. It could not have been a very demanding course, since he was back, degree in hand, within a few months. In any case, the degree provided backup if his career in botany ever fell apart.

By now, though, Linnaeus was fairly sure he would not need much backup.

PUTTING ORDER TO DIVERSITY

Linnaeus's adventures in Lapland, with its wide diversity of plant life, had brought him face to face with one of the oldest and most perplexing problems facing botanists and other natural historians. And he had found a way to solve it. Returning once again to Uppsala, he set to work on a small book that he had been planning, a book that he was sure would establish his credentials in botany once and for all.

He was right. Published in 1735, the slim volume entitled *Systema naturae* established a particularly methodical system of classification of living things and set him on his way to becoming the most famous scientist in Sweden, the founder of modern taxonomy. As further revised and expanded through the years, it finally gave botanists and naturalists a way to handle one of their most vexing and perplexing problems: how to name and classify the world's life forms. For as long as people had been observing and writing about nature, they had struggled with naming specimens and showing their relationships to other specimens. The same plant, for instance, might be found in widely different areas around the world—and might be called by a different name in each area. Natural scientists and students studying the plant needed to find a way to identify it by a name that they could all use when communicating about it. Ideally the name would also convey something descriptive and useful about the specimen that would allow them to group similar specimens together closely enough to study their similarities and differences. It was obvious, for instance, that a dog was different from a whale: whales lived in water and dogs did not. But how different was a dog from a cat? Both lived on land, both had fur and four feet. Dividing creatures into just two groups, those that lived on land and those that lived in water, wasn't helpful enough. The categories were too large and could be extremely misleading. In fact, a respected scientist of Linnaeus's time had followed just such a path and grouped beavers along with fish since both lived in water. Needless to say that was an extremely misleading classification—so much so that the Catholic church for a while permitted beavers to be eaten on fast days (when no meat was permitted, only fish)!

In the late 1600s the English naturalist John Ray had made a little progress in resolving the growing classification problems. But in the wake of a virtual flood of new specimens of plants and animals coming in from all over the globe as explorers and merchants continued to expand their operations, it was obvious that some new system for naming and classifying specimens was sorely needed. Only then could any serious scientific work be done on the exciting new finds.

Linnaeus knew that the system he came up with wasn't "natural," wasn't the ultimate plan of nature and didn't reflect the ultimate delineation of the Great Chain. But his chief aim was to create a system of nomenclature that was practical and easy to use. This he did with such success that much of his system, modified, is still used today, more than 200 years later. Linnaeus grouped plants and animals into groups he called genera (the plural of *genus*), and subdivided those groups into species. And he used what scientists like to call a *binary nomenclature*, a two-part (binary) system of names. Because Latin was still the international scientific language of the time, the names he gave were Latin or Latinized, and that tradition has stuck to this day. Each two-part name begins with the genus (the larger, more inclusive group), followed by the species (the smaller, more specific group). He grouped animals or plants together in one genus that seemed to have something in common (usually a structure, a body shape, or some distinctive way of reproducing, for instance). For example zebras and horses clearly have similarities, while horses and dogs have much less in common. So Linnaeus put zebras and horses in the same genus, *Equus*. Dogs, by contrast, are classified in the genus *Canis* (the same genus, by the way, as wolves). The shared genus name points to similarities in a group, but zebras and horses are also distinct from each other. So the zebra is named *Equus zebra*, while the horse goes by the name *Equus caballus*, the second part of the name, the species, emphasizing the unique aspect of different members of the genus. The naming process, Linnaeus admitted, was tricky. Arguments raged about what constituted a species: organisms that can reproduce with each other?

Linnaeus in Lapp costume (National Library of Medicine)

What is to be done with organisms that cannot reproduce, such as mules? "The first step in wisdom," wrote Linnaeus, "is to know the things themselves; this notion consists in having a sure idea of the objects; objects are distinguished and known by classifying them methodically and giving them appropriate names . . . classification and name-giving will be the foundation of our science." And to a great extent he was right. Without question his system went a long way toward bringing method to the study of living things. With the publication of *Systema naturae* Linnaeus became instantly famous. At last for many it seemed as if the logjam of classification had been broken by the Swedish naturalist. Linnaeus became a national hero, the Prince of Botanists, as many called him at the time.

Unlike many other scientific revolutions occurring during the 17th and 18th centuries, Linnaeus's great breakthrough had the misleading appearance of one without ideological violence. Unlike Copernicus and Galileo, he did not seem to be upsetting the Solar System or threatening established religious thought. On the contrary it seemed he was simply following in the footsteps of Adam, giving names to the creatures of God. He had given temporal order without disrupting religious order. The Great Chain of Being remained intact. His studies and classifying had convinced him that according to God's plan, the unchangeability of the species was the rule of nature. And each creature, with its secure place in the great chain, had been personally designed by the creator. While there is some evidence that later in his life Linnaeus may have begun to alter some of his views on new species slightly, he intended this, his most famous and influential work, to strongly support the traditional view of the fixity of the species and God's unique role in the creation.

"As there are no new species," he wrote, "(1): as like always gives birth to like (2): as one in each species was at the beginning of the progeny (3): it is necessary to attribute this progenitorial unity to some omnipotent and omniscient Being, namely *God*, whose work is called *creation*. This is confirmed by the mechanism, the laws, principles, constitutions and sensations in every living individual."

Although he had declined to follow his father's footsteps into the ministry, his work he believed had allowed him to trace a more important path, one identified, as he wrote later, with "the very footprints of the creator." Always a complex individual of many contradictions, Linnaeus never abandoned either his immodesty or his piety.

So, for most, Linnaeus's new system came like an intellectual miracle. Out of the blue from an impoverished university in Sweden had arrived at last a way to come to grips with the long-standing problem of classification. He had brought order and method to the identification of the tremendous diversity of life. Almost immediately too it gave rise to a tremendous boom in worldwide specimen collecting—with Linnaeus himself directing many eager students to travel the world in search of new finds. It was dangerous work; some estimates claim that one out of every three of Linnaeus's student emissaries died in their search. But no longer would new and mysterious specimens be collected only to be tucked away and forgotten in back rooms or curiosity museums. Now every plant or animal could be labeled or "identified" and each new discovery could find its rightful niche to yield new and useful knowledge in the quickly expanding map of nature. And the numbers, thanks to voyages like Captain James Cook's, swelled quickly. Linnaeus knew and gave scientific names to 4,200 species of animals and 7,700 plants. Today the system has been gradually expanded to include 350,000 plants and more than a million animals.

Linnaeus's system was a heady achievement for a largely self-taught naturalist. On his return to Sweden in 1741, he entered medical practice and was appointed to a chair of medicine. A year later he moved into the chair of botany at Uppsala. In 1761 Linnaeus, the son of a poor Swedish minister, was appointed a member of the Swedish House of Nobles and given the right to go by the name Carl von Linné. Dying in 1778, he remained to the end a legend and a revered figure in his native land.

But not everyone agreed with Linnaeus's work. There were evident flaws and weaknesses, and many scientists and philosophers in the intellectually critical and skeptical 18th century were uneasy with its inconsistencies. Some of those critics began to set in motion a movement that would result, in the following century, in a revolution in biology that would be more disturbing for many even than the great Copernican revolution of the 17th century: the theory of evolution.

Carolus Linnaeus always fought against any idea of evolution—and preliminary rumblings of the idea had already begun to be heard. All species, he believed, had been created separately in the beginning and no new ones, he was sure, had ever formed after the creation. Nor had any ever become extinct.

Linnaeus's work was changed slightly by those who followed him, among them Georges Cuvier, who changed some of the details to make the system of nomenclature a more natural one emphasizing relationships. Although neither Cuvier nor Linnaeus intended it, the system of classification formulated by the rigidly conservative Swedish botanist ended up leading relentlessly toward Charles Darwin and his theory of evolution in the following century.

CHAPTER 7

BUFFON AND DIVERSITY IN NATURE

Among those who were skeptical about Linnaeus and his new classification system, Georges Louis Leclerc, comte de Buffon (1707–88), was the most critical and the most influential. Born in France in the same year as Linnaeus, Buffon was in many ways Linnaeus's exact opposite. Whereas Linnaeus was born poor and struggled for money most of his life, Buffon was raised by wealthy and well-educated parents. Where Linnaeus was pious (his critics often said that he wrote as if he had been present at the creation), Buffon was skeptical. Whereas Linnaeus was methodical and restrained, Buffon was intuitive and speculative. Linnaeus's enemies called him stuffy, rigid and self-righteous. Buffon's enemies called him a dandy and a playboy.

There were more basic differences, though, between the two men than social position and personality. The world for Linnaeus was the magnificent and perfect work of God. His own "modest role" in this perfect work, Linnaeus believed not too modestly, was to carry on the work of Adam—to assign identity and give names and by doing so to help us understand the wonderful, underlying order and purpose of God's universe.

Buffon's world was a world governed not by divine order and purpose but by "a system of laws, elements and forces," a Newtonian world functioning within the laws of nature, whose processes were ends in themselves rather than a part of a divine purpose or plan. A world, as Newton saw it, of motion and continuity.

Like his contemporary Voltaire, Buffon had spent some time in England, in his case driven there by a youthful indiscretion that resulted in a duel and temporary exile. Also like Voltaire, he was quickly attracted to the beautiful logic and impressive successes of the Newtonians. To get a better hold on the English language he translated one of Newton's works on calculus into

Buffon's estate, Montbard, near Dijon, where he spent summers reading, studying nature and writing his vast Natural History (Figuier: *Vies des savants,* 1870)

French, and pored over English books, first on physics, his earliest interest, and then on botany. By the time he returned home to France, he had picked up not only Newtonian ideas and ideals but also an English mannerism and style.

Buffon was determined to offer the world his own comprehensive view of the universe. It was a big ambition for the formerly pampered playboy. Fortunately, his work was recognized by 1739 by the French Academy of Science, to which he was elected an associate, and at the age of 32 he was appointed director of the royal garden, the Jardin du Roi (now the Jardin des Plantes). In this position, he was able to amass an impressive collection of specimens and made the garden into a fine research center.

AN ENCYCLOPEDIA OF NATURAL HISTORY

The Jardin du Roi provided grist for Buffon's intellectual mill, and he was gifted not only with intelligence and a well-honed writing style, but with enough self-knowledge to know that he needed more discipline. He spent winters in Paris, but summers he spent at his estate, Montbard, near Dijon, where he set up a spartan regime beginning at six every morning. Knowing his own tendency to sleep in, he awarded his trusted valet extra money for successfully getting him out of bed, and he interrupted his work only twice a day to have his hair dressed and powdered. It was a discipline he maintained for 50 years. Thomas Jefferson, who was invited to dine at Montbard when he was American ambassador to France, recalled,

> *It was Buffon's practice to remain in his study till dinner time, and receive no visitors under any pretense; but his house was open and his grounds, and a servant showed them very civilly, and invited all strangers and friends to remain to dine. We saw Buffon in the garden, but carefully avoided him; but we dined with him, and he proved himself then, as he always did, a man of extraordinary powers in conversation.*

Buffon began writing his encyclopedic *Histoire naturelle* (*Natural History*) in 1745. The first three volumes were published in 1749 and immediately met with great success. Although he had originally planned to take only a few years and fill up just a couple of volumes, it was a project that occupied the rest of his long life and eventually filled 36 books (with the help of others) published during his lifetime (plus eight more published after his death).

Despite his tremendous discipline when it came to his writing, however, Buffon was not a very disciplined scientist. Although he was enamored of the Newtonian revolution, he was not really inclined to the kind of tough-minded observation, experimentation and mathematical analysis that was required to perform well in the scientific arena. But what Buffon tried to accomplish was a kind of overview of the universe in which he would peek into all of its nooks and crannies and, using Newtonian ideas and the new mechanistic viewpoint, offer his own speculations on what it all was and how it all came to be.

Despite its modernity, this exciting endeavor harked in many ways back to the ancients, who, Buffon pointed out, were ". . . great men and were not limited to a single field of study. They had lofty minds, wide and profound knowledge, and broad views." As if foreseeing criticism—and perhaps taking a stab at the detail-oriented Linnaeus—he continued, ". . . If, at first glance, they seem to us to lack exactitude in certain details, it is easy to see, on reading them with reflection, they did not believe that little things merit as much attention as we have recently given them."

Needless to say, with Buffon's thinking and writing stretched out over more than 40 volumes and nearly 50 years, today's readers can find many loose ends, contradictions, errors and instances of sloppy thinking in his momentous work. This said, then, it's important to recognize that his *Natural History*, in its time, was tremendously exciting and influential. Its volumes stimulated other, more disciplined thinkers to grab hold of some of Buffon's provocative and speculative ideas and focus their own more exacting researches.

The vast and fascinating collection of plant specimens amassed by Buffon in the Jardin du Roi made it a favorite visiting place for foreign dignitaries. (Figuier: *Vies des savants,* 1870)

In his first volume Buffon immediately established his differences with the Linnaean system, which he scorned as dry-as-dust compartmentalization, and with all "artificial" systems in general. "The error," he argued, "consists in a failure to understand nature's processes, which always take place by graduations. . . . It is possible to descend by almost insensible degrees from the most perfect creature to the most formless matter . . . there will be found a great number of intermediate species, and of objects belonging half to one class and half to another. Objects of this sort, to which it is impossible to assign a place, necessarily render vain the attempt to a universal system."

What, then, was Buffon proposing? Here the going gets tricky, since his ideas, always speculative, were in constant development and were spread out over many years and many books. His starting point was his belief that all classification systems were simply convenient products of the human mind and that nature itself was not made up of such discontinuous divisions as classes, orders, genera or species. No matter how helpful (or harmful) such categories might be to students of nature, they were purely artificial and arbitrary. Nature, said Buffon, is composed of individual organisms that show very small and continuous gradations one from the other. His ideas about classification apparently changed, though, over the years. In 1749, when his publications first began, he was strongly skeptical that any classification system was possible of the diverse world of living things. By 1755 he admitted that there were related species. Species, though, he argued, were "nature's only objective and fundamental realities." All other divisions remained artificial and misleading.

If he had stopped here, we would probably know Buffon today only as a mildly interesting footnote to 18th-century science history. But unlike those natural historians he scorned for writing books "filled with numerous and dry nomenclature" and "wearisome and unnatural methods," Buffon aimed to throw a wider net, to present a lively and evocative picture of nature as a whole, to offer a wide-ranging and general history of the Earth as a home to living creatures—a magnificent machine in motion. And motion was the clue. For Buffon was beginning to believe that perhaps life itself was a part of the grand movement—that, like the Earth and the physical universe that was its home, life, too, was not static but evolutionary.

THE INTERNAL MOLD

As described in Chapter 3, Buffon had some rather amazing ideas about the longevity of the Earth's history, even though he usually tried to stay out of trouble with the church authorities, who disapproved of suggestions that the Earth might be older than the 6,000 years suggested by the Bible.

Life, though, was what interested Buffon the most, and once setting the stage for it by his description of the history of the Earth, he began pursuing his most seminal and interesting ideas about the development of life forms on Earth. Although the development of those ideas, spread out over many books and many years, was often erratic and sometimes contradictory, their essence suggested the beginning of a major intellectual shift, one that would move beyond the static view of life and the intellectual bonds of the Great Chain of Being and set the stage for the great work of Charles Darwin in the 19th century.

Buffon began his speculations about life and its development with one strong belief in mind: that life, like the universe, could only be explained and properly understood if the explanation could be done in strictly mechanistic—that is to say, Newtonian—terms. He looked for physical explanations and cause-and-effect relationships. With his work on gravity, Newton had shown that such physical relationships were not always obvious. Objects did not have to be touching each other to affect each other. They didn't even have to be close: consider the effect of the Sun and the Moon on the Earth. In the life sciences, Buffon believed, this same premise could be followed.

Buffon was not averse to fudging his speculations occasionally when he ran into a dead end. He began with the long-standing problem of the embryo (which will be discussed in Chapter 8, on biological reproduction, or *generation*, as it was called in the 18th century). Buffon adopted the theory, held by some, that the embryo was formed from a mixture of male and female semen in the womb. (Though it sounds strange today, the idea of female semen was common at the time.) The semen, he theorized, was made up of "organic particles" derived either (depending on the stage of his thinking) from food, or from the atmosphere, which was full of minute living particles. Once collected into sperm, he explained, the particles then arranged themselves into the complex structure of the embryo. How, though, do the particles "know" to form one particular species rather than another?

Here Buffon attempted, not too successfully, to put a new spin on an old idea, Plato's eternal "essences," the idea that ideal forms exist outside space and time, independent of particular manifestations. Each species, Buffon suggested, was associated with an "internal mold," which somehow directed the particles into their proper place. (What this "mold" was, how it came to be, and how it actually manipulated the particles into place, Buffon never satisfactorily explained.) As products of this internal mold, the species were "fixed." Each species was the definite and specific product of a particular mold, which had existed throughout the universe for that species since the beginning of time. The mold determined all the details of an organism, shaping disorganized organic particles into a living animal or plant having specific characteristics. Buffon even thought that these pre-existing molds would produce the same species on another planet, exactly the same as on Earth, if the correct temperature and other environmental requirements developed (a radical idea for the time). In his "internal mold" hypothesis, Buffon differed from other, even more radical materialists who maintained that the origin of all life began as a result of spontaneous generation—at random, governed entirely by circumstances.

But Buffon also recognized, despite his attachment to this concept of immutability, that species did seem to change. During the course of his observations, he noticed, for example, the existence of rudimentary organs that were imperfectly formed and had no apparent purpose. "The pig," he wrote, "does not appear to have been formed upon an original special and perfect plan, since it is a compound of other animals: it has evidently useless parts, or rather parts of which it cannot make any use, toes all of the bones of which are perfectly formed and which nevertheless are of no service to it." These "useless parts" seemed to suggest the existence of imperfections in the species. If every species was perfectly formed from the beginning of time and had never changed since then, why would these imperfections exist now? This, he explained, was a process he called "degeneration," in response to the environment. But Buffon also believed that a species would revert back to its original form once the environmental influences that caused the degeneration were removed. Here again he differed from later theories of evolution, which would see these changes as part of a continuing flow of change, not as temporary or reversible. (Buffon's ideas, by the way, caused a stir even in far-off America, prompting Thomas Jefferson to send him the

Georges Louis Leclerc, comte de Buffon
(Figuier: *Vies des savants*, 1870)

skin of a large American panther to refute Buffon's claim that New World animals had degenerated in size.)

BUFFON'S LEGACY

Buffon was provocative, his writing was elegant and his influence as a popularizer of science was broad, and he had a considerable impact, not only on young zoologists of his day, but on the populace as a whole. Most of his work consisted of superbly written monographs on individual mammal species notable both for their scientific and their literary value. While many of his ideas shot wide of the mark, he stimulated thinking with his broad brush strokes where Linnaeus had plunged deep into the details. He resisted the idea of evolution of the species, steadfastly denying any such possibility; yet the facts he assembled seemed to point away from the doctrine of "fixed species" he supported. And, in fact, Charles Darwin considered Buffon to be "the first author who in modern times has treated [the origin of species] in a scientific spirit."

Buffon's double talk may have been an attempt to avoid trouble with the church authorities; many of his more radical contemporaries thought he equivocated too much. Or perhaps his work just reflected his own changes in opinion over the long period of time during which he wrote his books. In any case, he did not escape the church's ire entirely. On June 15, 1751, Buffon was called before the Faculty of Theology at the Sorbonne and instructed to withdraw certain parts of *Natural History*, which, according to his critics, contradicted dogma. Specifically, the authorities cited those portions concerning the age of the Earth, the birth of the planets from the Sun and his contention that truth can only be arrived at through science. Buffon promised to subdue the offending heresies and to tone down his writing in the future. But he continued to write provocatively, if cautiously. Buffon died in 1788. More because he was part of the aristocracy than because of his controversial ideas, revolutionaries vandalized Buffon's grave and the monument that had been raised to him. But his spirit inspired several great natural historians of his time, including Jean-Baptiste de Monet de Lamarck and Georges Cuvier, both of whom will be discussed in Chapter 9.

CHAPTER 8

THE ANIMAL MACHINE: PHYSIOLOGY, REPRODUCTION AND EMBRYOLOGY

Buffon, of course, was not alone in attracting accusations of heresy. No one in the 18th century exemplified the materialist doctrine of the post-Newtonian era more than Julien Offroy de La Mettrie, author of *L'Homme machine* ("Man, the Machine"), a little book published in 1748. And probably no one more completely outraged the church authorities (except, perhaps, Voltaire).

La Mettrie had a long history of getting in trouble with authorities. Early in his career he had translated into French the works of Herman Boerhaave, a widely respected Dutch teacher and writer in medicine and chemistry, and incurred the wrath of the medical faculty at the University of Paris (who were so conservative that they still had not accepted the works of Vesalius and Harvey from the 16th and 17th centuries). An incorrigible renegade, La Mettrie continued to get himself into trouble by publishing satirical pamphlets about the faculty and numerous heretical books. Though he himself was a priest, La Mettrie seemed to have no problem with setting forth materialist ideas that ran counter to religious dogma.

L'Homme machine, La Mettrie's most famous work, was published in Leiden (in the Netherlands, which was more liberal than France), probably in hopes of slipping it by the conservative French church authorities. In it, he described man as a machine governed completely by physical and chemical agents. It was a radical idea, the ultimate application of the Newtonian revolution to humanity itself. He also denied the dualism established by Descartes, which emphasized a dichotomy, or division, between mind, or soul, and body. A human being, La Mettrie maintained, can be seen as just another animal, a sort of "talking monkey." With this

The influx of plant and animal specimens into Europe, as well as the exchange of ideas throughout the 18th century, contributed greatly to advances in understanding about the process and nature of life. Here Bernard de Jussieu imports a cedar of Lebanon, which came to France via England. Jussieu, a professor of botany at the Jardin du Roi in Paris, first recognized that sea anemones are animals (not plants). (Figuier, *Vies des savants*, 1870)

book, La Mettrie set the stage like no one before him for modern biology, by looking at humans as not essentially different from animals.

Inspired by La Mettrie—who was more a polemicist and philosopher than a scientist—and the work of William Harvey before him, 18th-century scientists set about systematically exploring the machinery of all biological organisms. In the process three main areas of thrust emerged in the battle for understanding: the effort to understand the physiological mechanisms (including digestion and respiration) that make living organisms work, the reproductive process and embryo development.

For the post-Newtonian biologists, the Body Machine was the idea of the age. Harvey had shown with his study of circulation that blood coursed through arteries and veins like water through pipes, controlled by valves and pumped by the heart. And no one wanted to accept the idea that the analogy of body and machine could only go so far.

PHYSIOLOGY

Albrecht von Haller

One of the most outstanding biologists of the 18th century, Albrecht von Haller was born in Bern, Switzerland in 1708. Along with his many students (he was also one of the great teachers of the century), he explored the form and function of one organ after the other, establishing anatomy as an experimental science, and also applying dynamic principles to the study of physiology.

Haller was an exacting and painstaking experimentalist, whose quest to observe, note and know was so great that even on his deathbed he remained a scientist to his last breath. In his last moments, on December 12, 1777, surrounded by physicians and friends, he placed a finger on his wrist, felt the thready pulse wane and falter and reported dispassionately, "The artery no longer beats."

With equal determination, during his lifetime Haller explored the irritability of muscle and the sensibility of nerves, made important contributions to the physiology of the circulation, including circulation time and the automatic action of the heart, and gave the first solid discussion of respiration.

His *Elementa Physiologiae Corporis Humani* (1757–66) (*Elements of Physiology*) was known for being the last word of the century on the subject. The great 19th-century physiologist François Magendi once complained that whenever he thought he had conceived of a new experiment, he would find it already described in detail by Haller in the *Elements of Physiology*. Haller systematically extended the knowledge of anatomy, linked it to physiology by experiment and applied dynamic principles to physiological problems.

In vivisection experiments (done on living organisms) he demonstrated his painstaking methodology, a consistent, step-by-step approach to understanding function and process. He defined as "irritable" those parts that contracted when touched, and as "sensible" those parts that conveyed a message to the brain when stimulated. He tried various stimuli—pricking, pinching and certain chemicals. He tested tendons, bones, cerebral membrane, liver, spleen, kidney, all of which he found insensitive. And he found that irritability in muscles is caused by stimulation of nerves. For example, he could cause the diaphragm to contract by irritating several nerves in the area. And he explored the nature of contractile forces that occur spontaneously in muscle tissue of living animals or those that have just died.

Always his approach was analytical, objective, moving forward on the basis of data discovered. "As the nature of the brain and of the nerves is one and the same," he wrote regarding his study of the brain and nervous system, "so are these alike in function. In treating them we will so far as possible make use of our experiments, nor will we at first at least go beyond the testimony of our senses."

By experiment he found that only nerves serve as instruments of sensation and therefore only those parts of the body connected to the nervous system experience sensation.

Haller's work was unflagging and thorough, with experimental claims always substantiated by evidence. He could, of course, throw his weight behind a misleading theory, as will be explained later in this chapter, but he epitomized the biological experimenter of his time and, overall, his greatest contribution was to the spirit and method of physiological research. Though he is comparatively unknown today, his influence throughout Europe in his time and his general intellectual activity is said to be equaled only by Voltaire's.

René Antoine Réaumur

Considered one of the founders of entomology, René Antoine Réaumur (1683–1757) wrote a six-volume compendium on the life cycles and behavior of insects. His observations of living insects were so superb that they served to set the stage for *L'Histoire naturelle* by Buffon, his contemporary and rival. (In fact, Réaumur may have been behind the attack on Buffon made by the theological faculty of the Sorbonne.)

In Réaumur's work he also emphasized the larger groupings of insects, rather than getting into tedious description of species, thereby setting the stage for Cuvier and his modifications to Linnaeus's system of classification (see Chapter 9). Réaumur also invented a thermometric scale and a type of porcelain, and his researches helped establish the French tin-plate industry, as well as contributing greatly to steel-making in France. But in biology, his rare commitment to understanding the process of digestion is perhaps his greatest claim to fame.

He had a pet hawk of a type that vomits up indigestible foods, and Réaumur trained his pet to swallow small open-ended metal tubes containing sponges. When these tubes, predictably, came back up, he examined the sponges and found that, as the bird had attempted to digest them, they were soaked with gastric juice. Réaumur then took the gastric juice and showed that the juice, when applied to meat, tended to soften it. "When I put some of the juice from the buzzard's stomach on my tongue," he wrote, "it tasted salt rather than bitter, although the bones . . . on which the fluid had acted, had not a salt but a bitter taste." He also found that when he placed bits of meat in the tubes, they came back up partly digested—but not, as some theorists had suggested, either ground up or putrefied (rotted).

René Antoine Réaumur (1683–1757) studying the behavior of caterpillars (Figuier, *Vies des savants*, 1870)

Lazzaro Spallanzani

Lazzaro Spallanzani (1729–99) continued Réaumur's experiments, recognizing that when testing the effectiveness of gastric juices on meat, the temperature should be the same as the body temperature of the animal the juice came from. Spallanzani tried the same process with other birds, including a crow, by using string to pull the partially digested material back up. He found that after seven hours most foods were completely dissolved by the juices.

Not content with these results, though, Spallanzani went a step further, experimenting on himself. Though afraid that he might choke and die (as in fact, Réaumur's pet hawk had), he swallowed a small linen bag containing a few grains of chewed bread. When the bag emerged from his system 23 hours later, the bread was gone, although the bag was still in good shape. He later tried swallowing wooden spheres and open-ended metal tubes containing food, with the ends covered with gauze, but he gagged on them. His scientific curiosity, he had found, had run up against its limits on this particular subject.

Lazzaro Spallanzani, one of the greatest of all life sciences experimenters, performing an experiment on digestion in birds (Figuier, *Vies des savants*, 1870)

A trailblazer, the Italian physiologist Giovanni Morgagni (1682–1771) began looking at illness in a new way, by examining the anatomy of diseased, rather than healthy, tissue, interpreting the causes and progress of disease from the anatomical standpoint. In 1761, at the age of 79, he established the science of pathology with the publication of a book on the 640 postmortem dissections he had done.
(Parke-Davis, Division of Warner-Lambert Company)

LAVOISIER CRACKS THE RESPIRATION PROBLEM

So much of progress in science depends on all the previous hurdles having been jumped so that when a person with a certain perspective, background and aptitude comes along, he or she has a clear track ahead. Antoine Lavoisier, who also did so much in other areas of chemistry, had the superb luck to come to the problem of respiration with all the necessary hurdles behind him: Harvey had explained the circulation of the blood; an understanding of the movements involved in respiration had been achieved; the microanatomy of the lungs had been done by Malpighi; and the chemistry of gases, thanks in part to his own efforts, was now far more in hand than ever before. By the end of the century, successful isolation and characterization of the gases had produced a revolution in chemistry (see Chapter 4) and now, at last, respiration began to fall into place.

By 1777 Lavoisier had published a paper entitled "Experiments on the Respiration of Animals and on the Changes which the Air undergoes in passing through the lungs." Now that he understood the nature of the gases involved—"eminently respirable" air (oxygen) and "fixed air" (carbon dioxide)—Lavoisier could explain that respiration was a slow combustion or

EDWARD JENNER: VANQUISHING SMALLPOX

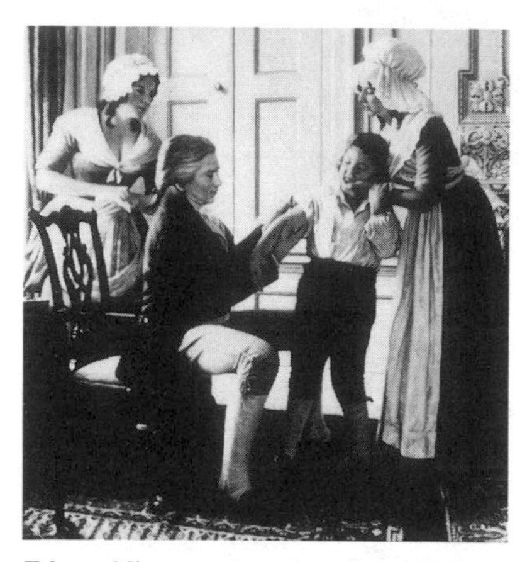

Edward Jenner administering a smallpox vaccination. (Parke-Davis, Division of Warner-Lambert Company)

While Edward Jenner (1749–1823) was still a young medical apprentice, he began thinking about a remark he had heard from a country dairymaid. The girl argued that she never had to worry about contracting smallpox since she had already once contracted cowpox. Cowpox was an extremely mild disease that was quite common and was transferred from the udder of a cow to the hands of its milker, resulting in pustular eruptions similar to (but much less serious than) smallpox. The fact that contracting cowpox led to immunity to smallpox was common knowledge in the rural Gloucestershire area where Jenner practiced as a country doctor.

It was already widely known that if a person survived a mild form of smallpox he or she was immune from further infection, and in fact some physicians were already inoculating some wealthy patients with a mild form of the disease to safeguard them against the great smallpox epidemics that swept in waves across the face of 18th-century Europe. The inoculation was expensive and the injection was almost as dangerous as the disease itself. It sometimes proved fatal, and it usually left the patient covered with ugly scars.

Jenner spent nearly 20 years studying the problem while keeping careful

oxidation. Although his English colleague, Joseph Priestley, had missed the point, Lavoisier grasped that respiration used oxygen and released carbon dioxide. He collaborated with Pierre Simon Laplace (see Chapter 2) to design an experimental device that could provide a quantitative measurement of the production of animal heat, in order to investigate the physiochemical basis of respiration. Using the ice-calorimeter they developed, they could now compare respiration and combustion quantitatively, with exciting results. No longer was the comparison of respiration and combustion just a nice metaphor. Lavoisier was now able to conclude in his "Memoir notes on cowpox and smallpox patients. Finally in May 1796 he ran his experiment. Using the cowpox pus from the hand of an infected dairymaid, he "vaccinated" an eight-year-old boy named James Phipps. The boy contracted a mild case of cowpox, as Jenner had expected, but quickly recovered, also as Jenner had hoped. The next step came two months later when Jenner inoculated the boy with what should have been a lethal dose of smallpox. It was a dangerous and controversial experiment, but the boy remained healthy and showed no signs of being infected with the deadly disease. Jenner repeated his test a few months later, again injecting young Phipps with yet another strong dose of the smallpox serum. Again the boy remained healthy.

Jenner concluded that the cowpox virus was similar enough to the smallpox that the body could build a resistance to both at the same time, but the cowpox disease was so mild that being inoculated with it caused only minor discomfort.

Coining the word *vaccination* from the Latin word for "cowpox" (*vaccinia* from the Latin for "cow," *vacca*), Jenner published his findings. Although there was skepticism and some resistance at first, vaccination was completely accepted in England by 1800 and was soon being adopted by the rest of the world. At one point Jenner himself was vaccinating more than 300 poor people a day in the garden of his country home. By 1800 an estimated 100,000 persons had received the new immunization, and many countries soon made vaccination mandatory, after which the number of cases of smallpox dropped dramatically.

Although it was later discovered that the vaccination against smallpox did not last for a lifetime but had to be "refreshed" or "boosted" later, Jenner's treatment not only helped to rid the world of a dread disease but established the science of immunization and opened the door to the research by Pasteur, Koch and others looking for cures and immunizations against other diseases.

on Heat" that "the heat released in the conversion of pure air to fixed air by respiration is the principal cause of the maintenance of animal heat."

HOW DO ORGANISMS REPRODUCE?

One of the toughest problems facing those interested in natural history was the question about how animals came to have offspring. First of all, the ovum

THE SPONTANEOUS GENERATION CONTROVERSY IN THE EIGHTEENTH CENTURY

The idea that some organisms may originate spontaneously from inorganic matter has come and gone many times—and, at a molecular level, has played an important part in several widely respected 20th-century theories of the origin of life.

In the 17th century Francesco Redi put to rest the idea, fostered by the Greek thinker Aristotle, that organisms (such as maggots, tapeworms or insects) could suddenly spring from slime or a piece of aging meat or rotting flesh or excreta. But, with the discovery of microscopic organisms in the 17th century, the question of spontaneous generation rose again (although their discoverer, Antony van Leeuwenhoek, thought that they came from parents like themselves).

In the 18th century, Buffon threw his rather significant weight behind the idea of spontaneous generation. A friend of his, English microscopist John Needham (1713–81), collaborated with him on a series of experiments in 1748 that seemed conclusive. Needham boiled mutton broth and sealed it in glass containers. A few days later, when he opened the containers, he found numerous microorganisms present. His conclusion: that the microorganisms had been produced from nonliving matter (the boiled, and presumably sterile, mutton broth). "There is a vegetative Force

(or egg) in the mammal was not even discovered until 1828, and no one had ever observed the union of an egg nucleus and a sperm until the end of the 19th century. Those key discoveries required techniques and equipment not available to 18th-century scientists. As a result, many of their ideas were what we would call shots in the dark. And they had a lot of them, sometimes based on evidence, sometimes based on intuition and sometimes based on their favorite world view. Some naturalists thought that the egg (for example, in chickens or frogs) had nothing to do with fertilization and the growth of the embryo. Many others either denied the existence of sperm or classified them as parasites, useless if not harmful to the reproduction process. Many, including Linnaeus, thought that external fertilization never occurred; Linnaeus flatly asserted, "Never, in any living body, does fecundation or impregnation of eggs take place outside the body of the mother." Some, though not all, of these questions were addressed by the excellent experimental work done by Lazzaro Spallanzani.

Spallanzani's passion for science was probably encouraged by his cousin Laura Bassi (1711–78), who held a chair in mathematics at the University in every microscopical Point of Matter," he deduced, "and every visible Filament of which the whole animal or vegetable Texture consists . . ." In line with suggestions made by mathematician-philosopher Gottfried Wilhelm Leibniz (1646–1716) of the existence of "monads" or living molecules, Needham thought that animals and plants decomposed, when they died, to "a kind of universal *Semen*," a "Source of all," from which new living matter arises again and again.

But 20 years later Lazzaro Spallanzani repeated the experiment, this time using more scientific controls. He boiled different flasks for different lengths of time and discovered that some types of microbes were more resistant to heat than others. Some died after slight heating, but others survived boiling for as long as an hour. In Needham's experiment, he pointed out, some spores present in the original mutton broth had survived the short boiling period used by Needham.

Spallanzani's experiments certainly proved the necessity of thorough sterilizing techniques, and, for the moment, they laid to rest the idea of spontaneous generation. But in 1810 the French chemist Joseph Louis Gay-Lussac argued that Spallanzani's sterile vessels had lacked oxygen—which would be necessary for spontaneously generated life. This point called the conclusiveness of Spallanzani's experiments into question and left the question of spontaneous generation up in the air once again.

of Bologna, a rare distinction for a woman in her time. Spallanzani's interests ranged wide, from geology (he led an expedition to Mt. Etna during its eruption) to physiology to physics (he held the chair of physics and mathematics at the University of Reggio) to Greek to philosophy (both of which he taught). But, despite his divergent interests, Spallanzani is considered to be one of the greatest of all experimenters. Through his carefully controlled experiments he overcame—at least temporarily—the age-old belief in spontaneous generation (see box). And he found time to perform some very interesting experiments regarding reproduction.

One idea that had been set forth by several prominent 17th-century scientists, including Harvey and Fabricius, was that the active agent for fertilization was contained in the semen but was not part of it. In fact, they thought it was nonmaterial, an invisible power, like magnetic power, and they called it *aura seminalis.* Working with frogs, Spallanzani was able to prove that this idea of the aura seminalis was incorrect. He did experiments in which the female was killed while the eggs were being emitted. Eggs that were normally emitted and came in contact with semen developed as usual.

Those that were taken from the body of the female by dissection, that had never come in contact with the semen (though presumably nearby enough to have been exposed, theoretically, to any aura the semen might have) did not develop.

While these results looked good, Spallanzani decided to go one step further. He fashioned pairs of tight-fitting taffeta pants for the male frogs in the next experiment. Despite the strange costume, the frogs tried to mate as usual. But this time, no seminal fluid could reach the female or her eggs. The semen, and all it contained, was retained inside the pants. And though the female discharged many eggs, none of them developed. But when Spallanzani painted some of them with some of the fluid retained in the taffeta pants: Voilà! The eggs that were painted developed normally. Spallanzani also collected semen directly from the male's seminal vesicles and carefully applied them to eggs. Those eggs that were treated in this way developed into tadpoles. One of the techniques of sound scientific experimentation involves keeping "controls" to make sure that the outcome being observed would not happen even if the experimenter had done nothing. So Spallanzani also included eggs in this experiment that were not treated—and these disintegrated. (If they had developed into tadpoles, too, he would have had to ask himself a lot of questions about the design of the experiment, about his procedures and about his basic assumptions. And he would have had to refigure his conclusions—or, more likely, reserve making any conclusions until he could run the experiment again.)

Spallanzani had come upon a method of artificial insemination. In 1779 he cinched the case when he successfully used a similar process on a female dog. Although he certainly was not the first to perform artificial insemination successfully (Arabian horse breeders had been doing it for centuries), Spallanzani was the first to perform the experiment scientifically, with careful controls and documentation of results.

He created quite a stir in the scientific community, to say the least, and in 1781 the Swiss naturalist Charles Bonnet (1720–93) wrote to him, predicting with some clarity, "I do not know, but one day what you have discovered may be applied in the human species to ends we little think of and with no light consequences."

But the question still remained which portion of the semen actually caused impregnation. So Bonnet suggested another experiment to Spallanzani, leading to another series of cliffhangers. Spallanzani placed a carefully measured quantity of semen on a watch glass. Then, using a small amount of gluten, a natural glue, he attached 26 eggs to another watch glass, turned it upside down and placed it over the watch glass containing the semen. The eggs were moist, and some of the semen apparently disappeared (presumably evaporated), but the eggs did not actually come in contact with the semen. When placed in water, these eggs did not develop. But when he painted the semen onto other eggs, they did develop, demonstrating that the semen was still potent. Once and for all, the aura seminalis was crossed off the list of suspects. From this Spallanzani concluded that "fecundation [or fertilization] in the fetid toad is not the effect of the aura seminalis, but of the sensible [perceivable with the senses] part of the seed."

Bonnet suggested testing whether other influences—blood, blood extracts, electricity, vinegar, wine, urine, lemon and lime juices, oils and more—might have interesting results. Spallanzani followed up on the suggestions, but none of these influences caused any development whatever. He tried to test how virile the semen was—what could take its ability to impregnate away—and he discovered several things that did not work, including dilution, exposure to a vacuum, cold and oil. However, semen did not fare well when exposed to heat, evaporation or wine, or when passed through a filter paper.

This last piece of information gave Spallanzani an idea for another experiment, in which he filtered the semen. The result was a thin liquid that had no power to fertilize eggs. And, in the filter paper, a thick, gummy residue remained. When Spallanzani tried painting this residue on the eggs, again the eggs developed. But for some reason Spallanzani ignored the significance of this important detail, coming to the conclusion that it was not the sperm left in the filter paper but the small portion of seminal fluid that was left behind that fertilized the eggs. So Spallanzani should have concluded correctly from his experiment that it is the sperm, not the fluid around the sperm, that does the deed. But for some reason that perhaps no one will ever understand, he did not reach this conclusion. He had some previous experimental results that seemed to cloud the issue—seminal fluid that he thought was free of sperm that fertilized, seminal fluid that contained sperm apparently killed by exposure to urine or vinegar that had worked. Or maybe he was thrown off course by his belief that sperm were parasites passed on from generation to generation through intercourse, a sort of universal venereal disease.

Meanwhile, the question of how an animal develops once an egg is fertilized was still very much up in the air.

WOLFF CHALLENGES PREFORMATION

"The ovary of an ancestress," Haller once wrote, "will contain not only her daughter, but also her granddaughter, her great-grand-daughter, and her great-great-grand-daughter, and if it is once proved that an ovary can contain many generations, there is no absurdity in saying that it contains them all." Known as preformation, this idea had been around a long time—although, depending on which theory of reproduction one sub-

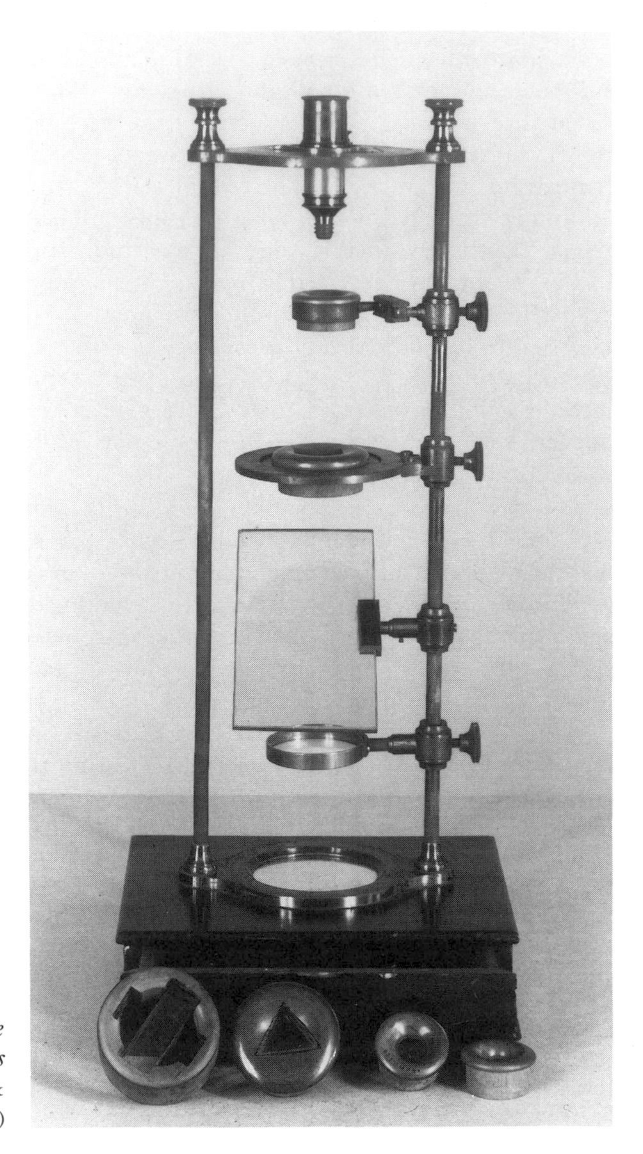

A microscope of the type used in the 1700s
(Courtesy of Bausch & Lomb)

scribed to, the "homunculus," as this tiny preformed human was called, might be located either in the ovum or in the sperm. And imaginative microscopists had even drawn tiny little men they thought they had seen huddled inside the confines of a single sperm cell.

But Caspar Friedrich Wolff (1734–94) brought a new approach to the analysis of embryo development and differentiation. In 1759 he wrote a landmark dissertation entitled *Theoria generationes* ("Theory of Generation"), in which he changed the course of the history of embryology. In it he recounted observations he had made on plants and outlined the philosophical premise that embryonic development was "epigenetic," that is, progressed through the gradual addition of parts.

Apparently unaware that most of his colleagues—including the great physiologist Albrecht von Haller—subscribed adamantly to the concept of preformation, Wolff sent his thesis to Haller (Wolff was only 26 at the time). Haller quickly rejected Wolff's thesis on religious grounds. No doubt feeling rebuffed, Wolff replied pointedly that a scientist must search for truth, without prejudging material on religious, not scientific grounds. For his part, though, Wolff was just setting forth a relatively unsubstantiated hypothesis; so his thesis, at this point, lacked the clout to stand up against Haller's objection.

But 18th-century biologists had come up against a dilemma. Although Haller, Spallanzani and Bonnet all felt comfortable with the religious framework and wanted to maintain allegiance to the biblical story of creation, they wanted to think of biological reproduction in the same mechanistic terms that they used for all the other aspects of physiology on which they were all working.

Using the experimental approach to crack this case posed problems. Microscopes had not improved particularly since the days of van Leeuwenhoek. Staining techniques, which would enable biologists to differentiate various kinds of tissue and cells, had not yet developed. More progress had been made with plants because details in plants are easier to see through a microscope than unstained animal tissues. And so, most of Wolff's conclusions for both plants and animals were based on his microscopic examination of plants alone. There, in the seed, he saw in the primordial undifferentiated material the rudiments of leaves and parts of flowers.

Then in 1768 Wolff described new researches he had succeeded in doing on the formation of the intestine of the chick. He had found that the chick intestine developed from a simple sheet of tissue as the embryo developed. It folded along its length to form a trough and then closed to form a tube. He also discovered in higher animals that an embryonic kidney develops that then later disappears before development is complete. (In his honor, the structure is now known as the Wolffian body.) In these two cases, he maintained, preformation theory could not hold up. The chick intestine could not have been preformed in the egg; he had watched it develop stage by stage. The process, he concluded, must be differentiation of simple uniform (homogeneous) tissue into more complex, heterogeneous (different) structures.

"We may conclude," Wolff wrote, "that the organs of the body have not always existed, but have been formed successively; no matter how this formation had been brought about. I do not say it has been brought about by a fortuitous combination of particles, by a kind of fermentation, through mechanical causes or through activity of the soul, *but only that it has been brought about.*"

Wolff was one of the first biologists to become a proponent of *Naturphilosophie* (nature philosophy), a German philosophy influenced by Romanticism, which held that nature was permeated by a life force that infused the process of creation and reproduction.

As a subscriber to *Naturphilosophie*, he thought the process he had observed exhibited a kind of life force that worked upon homogeneous organic matter, building it up into structures. But he could go no further.

In the end, although Wolff had gone far (T. H. Huxley considered him a greatly underrated genius), he could not win the battle against the theory of preformation. For that battle scientists needed the development of cell theory to provide a framework for understanding the sperm, egg and embryo. And, secondly, biologists would have to give up two of their favorite 18th-century ideas—both mechanism and *Naturphilosophie.* They had to give up trying to see organisms as machines, while not giving up the hope of explaining the mechanisms that make them work. Those who fought to understand the process of embryo development, like those struggling to grapple with the great questions of evolution, would only see their battle won in the coming era of the 19th century.

C H A P T E R 9

PRECURSORS TO MODERN EVOLUTIONARY THEORY: LAMARCK AND CUVIER

The 18th-century struggle over "fixity of the species"—the idea that all the species remained exactly as they always were—came to a climax at the end of the century in the work of two men, Jean-Baptiste de Monet de Lamarck (who is known as Lamarck) and Georges Cuvier. The fact that they were both French probably was no accident; Paris by this time had become the center for biological studies. And the city itself sat in a vast basin, a burial ground containing hundreds of thousands of shells, corals, mammals and a great repository of fossils. The two men, Lamarck and Cuvier, became arch rivals, one an expert on invertebrates (animals having no backbones), the other an expert on vertebrates (those having backbones).

LAMARCK OPENS THE DOOR

Charles Darwin once called Lamarck "the first man whose conclusions on the subject [of where species come from] excited much attention . . . ," one who "did the eminent service of arousing attention to the probability of all change in the organic, as well as in the inorganic world, being the result of law, and not of miraculous interposition." His place in science history is often overshadowed by Darwin's own enormous contribution in the following century and by the ridicule Lamarck received from his colleagues in his own time. But Lamarck was among the first to have the courage to set forth the idea that species are not fixed, but transform from generation to generation and over time. He was a man of brilliant intuitions, although he sometimes extended his speculations too far beyond his scientific knowledge base—attempting to synthesize theories, not only about biology, but also

Pioneering evolutionist Jean Baptiste de Monet, chevalier de Lamarck (Courtesy, Burndy Library)

about physics and chemistry (in which he loudly opposed Lavoisier's breakthroughs).

Born the youngest in a family of 11 children, Jean-Baptiste Pierre Antoine de Monet (1744–1829) nonetheless acquired the family title, chevalier de Lamarck. But that was the extent of his good fortune, for no income went with the title. As a young man with no real prospects, he was sent to study with the Jesuits to become a priest (his aristocratic family recognizing no other means of self-support honorable, except a military career). But as soon as Lamarck's father died, with what meager inheritance he did acquire, he bought a horse and rode away from his future life in the clergy. At the age of 16 he joined the German army; he distinguished himself during the Seven Years' War and was promoted for bravery. But a chronic inflammation of the lymph nodes in his neck nipped that career in the bud at age 22. So, still seeking his fortune, Lamarck went to Paris, where he married several times; studied botany, medicine and music; and worked as a bank clerk.

During this period he became a friend of the philosopher Jean-Jacques Rousseau, with whom he went on long walks, discussing nature and natural history. He found he could support himself by writing, turning out a couple of lackluster books on medicine and an annual almanac that, year after year for 11 years provided totally unreliable weather predictions based on his own system of meteorology.

But Lamarck's career really began to take off when he went back to the interest in plant life that had first been kindled when he was stationed in the army along the Mediterranean coast. He wrote a book on French plant life (*Flore française*), which was published in 1778 and met with great success, going through several printings. It was the first field guide to be organized as a key to help readers identify species of French flowers, and it was widely used by botanists and casual hikers alike.

Aided by Buffon, who took an interest in him and hired him to tutor his son, Lamarck's career in natural history began. In 1781, he became an assistant in the botanical department of the king's natural history museum, a paid position that gave him the opportunity to travel widely, collecting botanical specimens for the museum. Then came the Revolution. The regime by which he and Buffon were employed came to an end, and the old Jardin du Roi in Paris became the new Museum of Natural History. But with the Revolution came a stroke of luck. In 1793, when the revolutionaries were casting about for someone to name to the lowly position of professor of insects, shells and worms (what we know today as invertebrate zoology) at the museum, Lamarck was chosen—despite his lack of any real background in the subject. (At about the same time his future rival Cuvier was named to the clearly more prestigious position of professor of vertebrate zoology.) Unflappable, however, Lamarck saw his opportunity and, at the age of nearly 50, plunged into the subject, producing prodigious results, including a seven-volume *History of Invertebrates*.

Lamarck succeeded in bringing order where Linnaeus had left a sort of indiscriminate hodgepodge. Lamarck differentiated the arachnids (spiders, ticks, mites and scorpions, all of which have eight legs) from the insects (which have only six). He established categories for crustaceans (crabs, lobsters, crayfish and the like) and echinoderms (those with external spines, such as starfish and sea urchins). Lamarck in fact established the words *invertebrate* and *vertebrate* and coined the word *biology*.

In his best work, *Zoological Philosophy*, published in 1809, Lamarck set forth two factors that he believed were important to the evolution of the species: first, that all creatures have a fundamental tendency to evolve into higher life forms and that nature itself tends toward an increasing complexity. And second, the idea for which he is best known and most criticized, the inheritance of acquired characteristics.

Lamarck thought that life was generated spontaneously from a gelatin or mucuslike material, aided in the process by heat or electricity, and he disagreed with those who thought that great floods and other catastrophes had marked the history of the plant and animal world. Living beings, he thought, had experienced gradual changes in their environment over vast

THE HUNTER BROTHERS: TALE OF A RUFFIAN SCIENTIST

William Hunter (1718–83) was building a reputation for himself in London as one of England's top physicians and anatomists; meanwhile, his younger brother John (1728–93) was building a different kind of reputation in Glasgow. In contrast to his serious-minded brother, John was considered a wild, heavy-drinking and barely literate lout by everyone who knew him. He drank too much, fought too much, swore too much and rode his horse too wildly over the Scottish countryside. His one virtue, the townspeople said, was that he could work with his hands and seemed to have a natural gift with tools in his job as a cabinetmaker.

Why in the world his brother William invited him to come to London and take a job as his assistant in his private anatomical laboratory was a mystery to everyone. And just about everyone thought that no good would come of it for either brother.

There was no denying John's skilled hands, though, and although he did not give up his wild living, he quickly progressed from performing dissections (while his brother William lectured) to becoming the supervisor of William's laboratory. William in fact was so pleased with his younger brother's performance (although still disturbed by his unruly habits) that he managed to get John permission to study at the Chelsea Military Hospital. Surprisingly, John breezed through with distinction and in 1754 he studied surgery at St. George's Hospital. After a year as resident surgeon he joined his older brother as a full partner in William's private anatomical school.

All was not smooth, though, between the brothers. John was still refusing to wear his appropriate wig and insisted upon being called "Jack." And he was not too happy when William talked him into entering Oxford. "They wanted to make an old woman out of me, trying to stuff me with Latin and Greek," he told his friends after he dropped out as soon as the first term finished.

He did manage to improve his reading and writing, although he never felt comfortable with either. And he did pick up a taste for good living, including the best seats at the theater and moving in London's elite social circles. More important, he quickly became known as one of London's most important researchers into the basic aspects of biology. He was also rapidly passing up his older brother, not only in research—he traced the development of the testes in the male fetus, followed the nasal and olfactory cranial nerves and studied the formation of pus—but in his reputation as an anatomist.

Although he pioneered in many areas, John Hunter is best remembered for the extensive museum of comparative anatomy he assembled. He began the collection to illustrate comparative structure and function of organs throughout the animal kingdom and spent long hours preparing the exhibits. The British government bought the extensive collection after Hunter's death and entrusted it to the care of the Royal College of Surgeons. Unfortunately much of it was damaged during World War II. (Parke-Davis, Division of Warner-Lambert Company).

William's reputation was firm. His specialty was obstetrics and, with his scientific methods, he had succeeded in greatly reducing London's mortality rate during childbirth. In many ways his had been a heroic effort, requiring a long fight to break the hold of midwifery in childbirth, a practice that had in the 17th century degenerated from a warm and caring procedure to an often careless one performed for quick profit by ill-trained practitioners.

Still, William probably was not too unhappy when his younger brother, concerned about his health, decided that he needed a warmer climate and joined the military as a surgeon. With his condition improved after a stint in Portugal—where he also spent time studying animal and plant specimens—John became staff surgeon on a military expedition to Belle Isle, a small island in the Atlantic. During fighting there between the British and Spanish troops he began an important study of gunshot wounds and developed his reputation as a surgeon still further.

Returning to London, John discovered that his former position at William's school had been filled, and he set up a private practice on his own. Like his older brother he also taught anatomy and surgery. And, despite his continued disdain for "book learning," he was the much more scientific of the two brothers, developing and taking special care in his procedures. Dissecting and studying more than 500 species of animals, John did pioneering work in the study of the lymphatic system, demonstrated that tendons would reunite after being severed, and made major investigations in the study of blood coagulation. He also worked on problems dealing with the embryology of the chick, and his preserved works in this area include some of the most beautiful and accurate drawings ever made on the subject.

Elected a fellow of the Royal Society of London in 1767, John managed to overcome his discomfort with writing often enough to make many major and varied contributions to the literature of science. His work demonstratred his agile and curious mind, ranging from "A Treatise of the Natural History of Human Teeth," to "Directions for Preserving Animals and Parts of Animals for Anatomical Investigations" to "Observations and Reflections on Geology."

Not surprisingly, during the latter part of his life, John found himself in many heated quarrels with his brother (each claimed the other had at various times "borrowed" from his work), and their disputes continued until William's death in 1783. John died 10 years later as a result of an experiment he had begun in 1767 when he inoculated himself with the pus of a syphilitic patient. Ironically his theories about syphilis and gonorrhea (too involved to go into here) were wrong and, beside misleading the study of syphilis for half a century, led to his death in 1793.

Characteristically, he studied the progression of the disease in himself until the end. Characteristically, too, he died following a heated quarrel with a colleague at a board meeting of the Royal Society. The immediate cause of his death was stated to be apoplexy, but more likely it was his untreated syphilis.

---

spans of time—in fact, Lamarck had a closer grasp than most of his contemporaries of the colossal length of geological time. These long-term changes in the environment, he thought, had not directly created changes in the species. But Lamarck proposed that changes in the environment affected the nervous system, which brought about changes in the entire structure of creatures: An urge, or a sense of need, within the animal somehow created changes in muscles and organs to respond to the changing demands of its surroundings. Lamarck thought that requirements of an animal's life actually shaped its organs and that these traits, when shared by male and female, were then passed on to the animal's offspring. For example, giraffes (which he believed began as antelope) stretched their legs, necks and tongues to reach the high leaves off the trees. Each generation succeeded in stretching a little farther, passing longer and longer legs, necks and tongues on to their offspring, until finally the giraffe as we know it today emerged. Other animal traits also evolved over time from generation to generation. Moles and blind mice, for example, cannot see because they live underground and have lost the sight of their eyes through disuse. Ducks' toes are webbed from stretching to paddle against the water. Lamarck did not say that species acquire new characteristics merely by wanting or desiring them, as many of his critics have wrongly claimed; his theory was much more mechanical than that. He thought that the urge or need within the organism translated itself into a kind of fluid, which flowed into the organ that needed to change—for example, when a giraffe stretched its neck to reach the leaves.

The "inheritance of acquired characteristics," however, does not hold up well to close examination. Giraffes might succeed in stretching their necks—a little. But what about camouflage coloration: stripes, splotches or spots? Most animals could not cause a change in coloration by trying or even needing to. And the evidence did not support the idea that acquired characteristics could be inherited, even if an individual did succeed in changing its anatomy.

But, despite getting sidetracked and being forever stigmatized on this point, Lamarck undeniably brought evolution to the forefront in biological thought, and he was the first biologist of any stature to do so. Erasmus Darwin (the grandfather of Charles Darwin) had put forth the idea some 50 years earlier, but his work was speculative and not really well thought through (a point that always embarrassed his grandson). And Cuvier refuted it flatly, preferring the idea of catastrophism (see Chapter 3 on geology). Over this point Lamarck quarreled bitterly with Cuvier—and Cuvier, a bad man to cross, never forgave that quarrel.

In his later years Lamarck became blind and, in a supremely low moment, Cuvier sarcastically chided him, "Perhaps your own refusal to use your eyes to look at nature properly has caused them to stop working." But Lamarck continued to attend scientific meetings on the arm of a daughter, trying to argue his point. "It is not enough to discover and prove a useful truth," he wrote, "it is also necessary to be able to get it recognized." In 1829 he died penniless, with his useful truth still unrecognized. Overall, the excitement Lamarck kindled in his contemporaries as well as those who followed was caused, unfortunately, more by derision than by scientific interest.

AND CUVIER TRIES TO CLOSE IT

Much of that disfavor and derision was thanks to the efforts of his arch-rival, Georges Cuvier, a giant of science in his time, whose formidable powers of observation and deduction won him well-earned acclaim, and whose political shrewdness gained him consistent success and position throughout the divergent regimes through which French politics passed in his lifetime. Because of the power of his stature, Cuvier's ideas totally eclipsed Lamarck's during his generation, as they did Hutton's in the field of geology (see Chapter 3).

Cuvier built his opposition to Lamarck's evolutionary ideas upon the solid foundation of his knowledge of comparative anatomy, which led him to think that the functions of all portions of an animal's anatomy and physiology were so completely integrated that no change could take place from generation to generation without disturbing the balance. Any part of an animal's body, Cuvier contended, was intrinsically related to all other parts of its body—with form following function. The shape and utility of one organ suggested to him a constellation of related organs and functions. And

Georges Cuvier, masterful comparative anatomist and paleontologist (Courtesy, Burndy Library)

he could reconstruct an entire animal life-style, literally in his sleep—or at least a famous story maintains that premise.

It seems that late one night (perhaps after a few glasses too many at the local café) one of his students determined to play a prank on Cuvier, appearing at his bedside dressed as a devil. "Cuvier, Cuvier, I have come to eat you," he growled menacingly. To which Cuvier, still half asleep, calmly replied, "All creatures with horns and hooves are herbivores. You can't eat me." And went back to sleep.

The son of a soldier retired from the French army, Georges Léopold Frédéric Dagobert, baron Cuvier, was born in 1769 in Montbéliard, a little town in Switzerland, to a relatively poor family of French Huguenots who had fled there for safety during the intolerant reign of Louis XIV. (Cuvier's Protestantism never stood in his way, though, in 18th-century France and the French revolutionaries annexed the area in which he was born in 1793, making Cuvier officially a French citizen.)

A child prodigy encouraged by his mother, Cuvier learned to read at four and entered the Academy of Stuttgart at 14, where his disciplined approach to his studies and his prodigious memory gained him recognition. (It was said that later in life he could recite from memory any paragraph from any book in his 19,000-book library.) As a boy, Cuvier became particularly fascinated with Buffon's books, which he found on the bookshelves of his uncle, who—like a great many Europeans—was collecting them, volume by volume, as they came out.

After his graduation at the age of 19, Cuvier became a tutor in Normandy to the 13-year-old son of a count. Not only did Cuvier become seriously interested in science at this point in his life, but he also gained many helpful social skills through his contact with a different segment of society, including counts, former generals and at least one friend of Voltaire. He also met a zoologist, Etienne Geoffroy St.-Hilaire, who in 1795 helped him secure the position of professor of vertebrate zoology at the Museum of Natural History in Paris. Napoléon, whose successes as a general in the revolutionary wars had won him acclaim, invited Cuvier to accompany him to Egypt in 1798. Napoléon liked Cuvier, and when he came to power, first as consul and then as emperor, he found positions for Cuvier in his government. In 1803 Cuvier became permanent secretary of physical and natural sciences of the Institut National, and in 1808 Napoleon put him in charge of investigating education in France. When the Bourbon royal family returned in 1815, Cuvier might well have lost out; but they used Cuvier instead, making him chancellor of the former Imperial University, now again the University of Paris. Cuvier also served in the cabinet of Louis XVIII, although, when Louis's more reactionary brother, Charles X, succeeded him in 1824, Cuvier had a brief fall from prominence. But by 1831, with Charles X exiled once more, the new king, Louis-Philippe, made him a baron. He also appointed him minister of the interior, although Cuvier did not live long enough to fill that position. He died during a cholera epidemic in May 1832.

Despite his ardent opposition to any hint of evolutionary theory, Cuvier contributed several ideas that made its formulation possible in the 19th century. He was the first to think of applying his principles of comparative anatomy to fossils. He recognized that fossils, entombed as they are in rock strata, represent a moment of time in the Earth's history, immobilized for examination. He urged his fellow scientists to ask questions, to examine, to engage in empirical study.

"Naturalists," he harangued, "seem to have scarcely any idea of the propriety of investigating facts before they construct their systems."

Collectors had amassed great accumulations of fossils, picked up willy-nilly, mindlessly assembled. Cuvier scolded them for regarding their fossil collections "as mere curiosities, rather than historical documents," for neglecting to look at what laws might govern the positions in which they were found or their relation with the strata in which they were found. He set forth a series of questions that he thought needed to be asked, such as "Are there certain animals and plants peculiar to certain strata and not found in others? What are the species that appear first in order, and those which succeed? Do these two kinds of species ever accompany one another?" And so on, in methodical order.

While we might be impressed by the rigor of Cuvier's approach, it is easy to be unimpressed by the fact that he thought he could find clues to the Earth's history in this way, since we take that idea so much for granted today. But, in order to establish a geological record from fossils, the ancient creatures of a particular species had to be found only in the rocks laid down in specific ages. For that to be true, the fossils had to represent extinct species, and that was a very hot issue in Cuvier's time. Many of his contemporaries (including Thomas Jefferson) thought that extinction could not possibly have occurred. Those fossils that represented species currently unknown, they maintained, must represent a species still living somewhere on Earth and as yet undiscovered.

Cuvier developed the ability to reconstruct a whole animal from a few parts. In 1796 he examined the ancient fossil of an elephantlike creature and found that it was neither of the two living species. He showed that the fossil of a South American animal, the *Megatherium*, was a giant ground sloth, now extinct, but related to smaller sloths in existence today. In 1812 he named a great flying reptile a *pterodactyl* because its membrane-wing stretched out along one enormous finger. (In Greek, *pteron* means "wing" and *daktylos* means "finger.") His explanation of these discoveries was that life's history must reflect a sequence of creations (and extinctions), each one more modern than the last.

Cuvier's knowledge of anatomy and bone structure of animals both living and extinct was unparalleled in his time. (The Bettmann Archive)

Cuvier also refined Linnaeus's system of classification, dividing the animal kingdom into four basic "types" and emphasizing the basically parallel nature of Linnaeus's arrangement. (The parallel ranking had serious consequences for the long-prevalent Great Chain of Being, not to mention Lamarck's hierarchical approach, both of which Cuvier rejected.) The four types represented groupings of animals with similar internal structures, four master plans with infinitely variable external structures adapted to the demands of the environment. Similarities of internal structures, not an ordered ranking of external characteristics, formed the basis for his classifications, and relationships between the species were based on these similarities. (Cuvier was also the first to include fossils in his system of classification.)

Cuvier's work with fossils and his system of classification set the stage, in a way, for Darwin's idea that in natural development a single original form can change in several different ways at once, leading to several kinds of diversification, with the ones best adapted to a specific environment being naturally perpetuated, or selected.

But, despite all this, Cuvier was against evolution. He knew that fossils must be ancient—buried as they were in rocky strata that could only have formed over a great passage of time. And he knew that the deeper the fossil and the older the rock, the greater the differences between the fossil and modern anatomical structures. He had set up a system of classification that implied diversification from an original model. Why then did he not make the easy leap to evolutionary theory?

Some critics say that Cuvier had a major blind spot: He believed in the biblical account of the Earth's history as recounted in Genesis. But Cuvier maintained that his objections were primarily scientific; based on his knowledge of the way the internal workings of animals fit together, he was sure that the species were fixed and distinct and that evolution was impossible. And he saw in catastrophism an idea he believed explained the existence of ancient fossils and apparently extinct species. Radical Enlightenment materialists, such as La Mettrie and Erasmus Darwin, had speculated about the origin and mutability of life. But the newly discovered facts of comparative anatomy made those concepts seem naive. The reality of biological complexity seemed to preclude the creation of these forms by a natural process.

By the time of Cuvier's death, the 19th-century geologist Charles Lyell had already made a few inroads on his position, though. He strengthened the case for the uniformitarian doctrine—the idea set forth by James Hutton that processes currently occurring on Earth provide a basis for understanding processes that have taken place in the past, the "steady-state" Earth—explaining all the Earth's features on this basis, without needing to resort to catastrophic theory. It was the beginning hint of a revolution to come in the history of biological thought.

Both Lamarck and Cuvier, ironically, are best known for being wrong. Lamarck, for his association with the idea of inheritance of acquired characteristics (which he did not originate and which was not central to his thinking). And Cuvier for his insistence on catastrophism and his belief in the fixity of the species. Both deserve better at the hands of the history of science.

Georges Cuvier was an exacting methodologist, and the brilliance, order and system he brought to his science produced an effect that far outshone his "mistakes." His dedication and showmanship brought respect and admiration to the search for understanding that science represents. Charles Lyell, the very man whose later publications in part produced the downfall of Cuvier's reputation, wrote this description of a visit he paid to Cuvier and his own recognition that the setting Cuvier had created to work in reflected his great involvement in his work, the kind of system and order that it required and the contribution it made to his prodigious output:

> *I got into Cuvier's sanctum sanctorum yesterday, and it is truly characteristic of the man. In every part it displays that extraordinary power of methodising which is the grand secret of the prodigious feats which he performs annually without appearing to give himself the least trouble. . . . There is first the museum of natural history opposite his house, and admirably arranged by himself, then the anatomy museum connected with his dwelling. In the latter is a library disposed in a suite of rooms, each containing works on one subject. There is one where there are all the works on ornithology, in another room all on ichthyology, in another osteology, in another law books! etc., etc. . . . The ordinary studio contains no bookshelves. It is a longish room, comfortably furnished, lighted from above, with eleven desks to stand to, and two low tables . . .*

Cuvier's empiricism—his insistence on looking at the results of observation and experiment—embodied everything that was exhilarating about the Age of Reason. And Lamarck ably pointed forward toward one of the greatest and most powerful theories of modern science. If about some things they were wrong, they only suffered the burden inevitable to anyone who risks asking questions and daring answers. As science writer Stephen Jay Gould has written on this subject, "Some types of truth may require pursuit on the straight and narrow, but the pathways to scientific insight are as winding and complex as the human mind." In that statement lie both the challenge and the wonder of science—that truths unfold in a strange and weaving manner—as the scientific adventures yet to come in the 19th century would further demonstrate.

EPILOGUE

Newton had planted the signpost and throughout the 18th century his road was heavily traveled. The scientific revolution had produced a great, new confidence in the human mind to solve all of its problems without recourse to mysticism or outdated authority—and that confidence had propelled the Western world through the major stresses of sociopolitical upheaval and technological change.

If the 18th century lacked giants the equal of Galileo or Newton, it did not lack in brilliant and persevering seekers who left deep and steady footprints along the road to unravel nature's mysteries.

By the turn of the century in fact, marching to the heady tune of Newton's laws, science seemed headed toward inevitable progress in every discipline. From Laplace to Kant, from Priestley, Cavendish and Black to Lavoisier, from Coulombe to Franklin, from Linnaeus to Buffon, Spallanzani, Wolff, Cuvier and Lamarck, the 18th-century thinkers went forward.

It was a heady and exhilarating time for scientific thinkers. And although the future would see some disturbing surprises and open up a new series of complex puzzles—many of which remain unsolved to this day—few thinkers doubted, by the end of the 18th century, that soon all of nature's secrets would be solved. If the universe was a magnificent machine (and after so much success had come from studying it as if it were, who could doubt it?), then it seemed the newly devised tools and methods of science required only the use of discipline and intelligence—and the human mind would soon understand all of nature's true workings.

Hubris, over-confidence, over-optimism? Perhaps. But, nonetheless, 18th-century science left a lasting gift to the future: a tradition that harnessed the problem-solving powers of the mind and gave those powers a permanent and central place in world culture, a tradition that would contribute to every aspect of human life, understanding and well-being in the century to come.

APPENDIX

THE SCIENTIFIC METHOD

. . . our eyes once opened, . . . we can never go back to the old outlook. . . . But in each revolution of scientific thought new words are set to the old music, and that which has gone before is not destroyed but refocused.
—A. S. Eddington

What is science? How is it different from other ways of thinking? And what are scientists like? How do they think and what do they mean when they talk about "doing science"?

Science isn't just test tubes or strange apparatus. And it's not just frog dissections or plant subkingdoms. Science is a way of thinking, a vital, ever-growing way of looking at the world. It is a way of discovering how the world works—a very particular way that uses a set of rules devised by scientists to help them also discover their own mistakes.

Everyone knows how easy it is to make a mistake about what you see or hear or perceive in any way. If you don't believe it, look at the two horizontal lines on the next page. One looks like a two-way arrow; the other has the arrow heads inverted. Which one do you think is longer (not including the arrow heads)? Now measure them both. Right, they are exactly the same length. Because it is so easy to go wrong in making observations and drawing conclusions, people developed a system, a *scientific method,* for asking "How can I be sure?" If you actually took the time to measure the two lines in our example, instead of just taking our word that both lines are the same length, then you were thinking like a scientist. You were testing your own observation. You were testing the information that we had given you that both lines "are exactly the same length." And, you were employing one of the strongest tools of science to do your test: you were quantifying, or measuring, the lines.

The Greek philosopher Aristotle told the world that when two objects of different weights were dropped from a height, the heaviest would hit the ground first. It was a common-sense argument. After all, anyone who wanted to try a test could make an "observation" and see that if you dropped

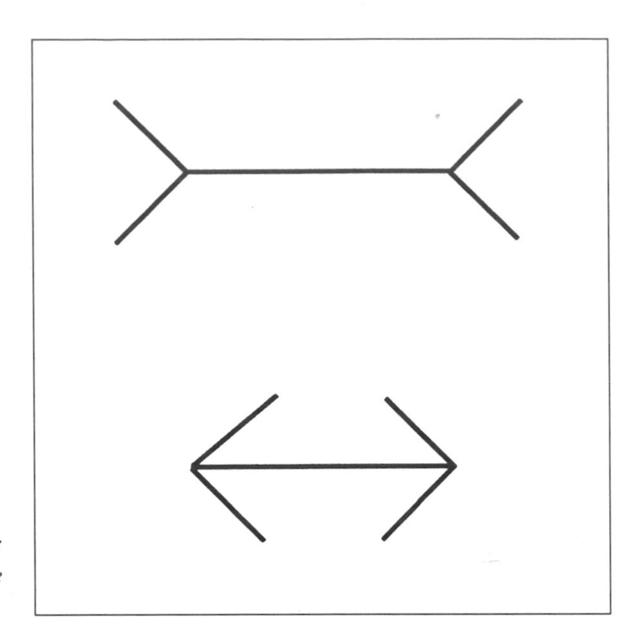

Looks can be deceiving: These two lines are the same length.

a leaf and a stone together the stone would land first. Try it yourself with a sheet of notebook paper and a paperweight in your living room. Not many Greeks tried such a test, though. Why bother when the answer was already known? And, being philosophers who believed in the power of the human mind to simply *reason* such things out without having to resort to *tests,* they felt that such an activity would be intellectually and socially unacceptable.

Centuries later though, Galileo Galilei, a brilliant Italian who liked to figure things out for himself, did run some tests. Galileo, like today's scientists, was not content merely to observe the objects falling. Using two balls of different weights, his own pulse as a time-keeping device, and an inclined plane, he carefully *measured* the movement of the balls down the ramp. And, he did this not once, but many times, inclining the plane at many different angles. His results, which still offend the common sense of many people today, demonstrated that, if you discount air resistance, all objects would hit the ground at the same time. In a perfect vacuum (which could not be created in Galileo's time) all objects would fall at the same rate! You can run a rough test of this yourself (although it is by no means a really accurate experiment) by crumpling the notebook paper into a ball and then dropping it at the same time as the paperweight.

Galileo's experiments (which he carefully recorded step by step) and his conclusions based on these experiments demonstrate another important attribute of science. Anyone who wanted to could duplicate the experiments, and either verify his results, or, by looking for flaws or errors in the experiments, prove him partially or wholly incorrect. No one ever proved Galileo wrong. And years later when it was possible to create a vacuum (even though his experiments had been accurate enough to win everybody over long before that), his conclusions passed the test.

Galileo had done more than show that Aristotle had been wrong. He demonstrated how, by observation, experiment and quantification, Aristotle, if he so wished, might have proven himself wrong—and thus changed his own opinion! Above all else the scientific way of thinking is a way to keep yourself from fooling yourself—or, from letting others fool you.

Of course science is more than observation, experimentation and presentation of results. No one today can read a newspaper or a magazine without becoming quickly aware of the fact that science is always bubbling with *theories*. "ASTRONOMER AT X OBSERVATORY HAS FOUND STARTLING NEW EVIDENCE THAT THROWS INTO QUESTION EINSTEIN'S THEORY OF RELATIVITY," says a magazine. "SCHOOL SYSTEM IN THE STATE OF Y CONDEMNS BOOKS THAT UNQUESTIONINGLY ACCEPT DARWIN'S THEORY OF EVOLUTION," proclaims a newspaper. "BIZARRE NEW RESULTS IN QUANTUM THEORY SAY THAT YOU MAY NOT EXIST!" shouts another paper. What is this thing called *theory?*

Few scientists pretend any more that they make use of the completely "detached" and objective "scientific method" proposed by Francis Bacon and others in the dawn of the scientific revolution. This method, in its simplest form, proposed that in attempting to answer the questions put forward by nature the investigator into nature's secrets must objectively and without preformed opinions observe, experiment and gather data about the phenomena. After Isaac Newton demonstrated the universal law of gravity, some curious thinkers suggested that he might have an idea *what gravity was.* But he did not see such speculation as part of his role as a scientist. "I make no hypotheses," he asserted firmly. Historians have noted that Newton apparently did have a couple of ideas, or hypotheses, as to the possible nature of gravity, but for the most part he pretty much kept them private. As far as Newton was concerned there had already been enough hypothesizing and too little attention paid to the careful gathering of testable facts and figures.

Today, though, we know that scientists do not always follow along the simple and neat pathways laid out by the trail guide called the scientific method. Sometimes, either before or after experiments, an idea or a hunch (that is, a somewhat less than well thought out hypothesis) suggests a new approach or a different way of looking at a problem to a scientist. Then he or she will run experiments and gather data to attempt to prove or disprove this hypothesis. One of the great differences between the hypothesis of the scientist and that of a non-scientist is that for a scientific hypothesis to be valid it must also be falsifiable. That is, it must have built within it some way that it can be proven right—or wrong, if it is wrong.

Not all scientists actually run experiments themselves. Most theoreticians, for instance, map out their arguments mathematically. But hypotheses, to be taken seriously by the scientific community, must always carry with them the seeds of falsifiability by experiment and observation.

If a hypothesis withstands the experiments and supplies the necessary facts to make it acceptable, not only to the scientist conducting the experiments or making the observations, but to others performing other experiments and observations as well, then when sufficiently reinforced by continual testing and appraising, the hypotheses may become known to the scientific or popular world as a *theory*.

It is important to remember, though, that a theory is also subject to falsification or correction. A good theory, for instance, will make predictions—events that its testers can look for as a further test of its validity. By the time most well-known theories such as Einstein's or Darwin's reach the textbook stage, they have run the gamut of verification to the extent that they have become productive working tools for other scientists—but in science, no theory can be accepted as completely proven but must remain always open to further tests and scrutiny as new facts or observations emerge. It is this insistently self-correcting nature of science that makes it both the most demanding and the most productive of humankind's attempts to understand the workings of nature. This kind of critical thinking is the key element of doing science.

The cartoon-version scientist portrayed as a bespectacled, rigid man in a white coat, certain of his own infallibility, could not be farther from reality. Scientists, both men and women, are as human as the rest of us—and they come in all races, sizes and appearances, with and without eyeglasses. As a group, because their methodology focuses so specifically on fallibility and critical thinking, they are probably even more aware than the rest of us of how easy it is to be wrong. But they like being right whenever possible, and they like working toward finding the right answers to questions. That is usually why they became scientists.

C H R O N O L O G Y

THE EIGHTEENTH CENTURY

1642	Disagreements between Charles I and Parliament lead to the English Civil War
1643	Louis XIV becomes king of France
1649	Charles I of England is executed
1654	Bishop James Ussher (1581–1656) publishes his argument based on biblical chronology that the Earth was created by God on October 26, 4004 B.C. at 9:00 A.M.
1654	The Black Death strikes eastern Europe
1662	The Royal Society is founded in England
1666	Paris Academy of Sciences founded
1676–78	Edmund Halley charts stars in the Southern Hemisphere
1687	Isaac Newton's *Principia mathematica* is published
1689	The English Parliament passes the Bill of Rights
1690	John Locke publishes his *Essay Concerning Human Understanding*
1691	Thomas Burnet publishes his "Sacred Theory of the Earth," an attempt to reconcile his belief in Newton's new "natural causes" philosophy with traditional biblical belief
1695	John Woodward's "Essay Towards a Natural History of the Earth" is published. Woodward believes fossils to be the remains of living things but attributes their demise to the biblical flood
1698	Sailing aboard the ship *Paramour Pink*, Edmund Halley begins first ocean voyage devoted solely to scientific research
1700	With a population of about 550,000, London is the largest city in Europe
1704	Newton's *Opticks* is published
1705	Halley predicts the return in 1758 of the comet that will eventually be called Halley's Comet
1709	New techniques in producing iron, which could be used in manufacturing better machinery, are developed in England, setting the stage for the Industrial Revolution

1718 Halley discovers proper motion of the stars

1727 James Bradley explains the aberration of the stars because of the Earth's motion

1727 Stephen Hales publishes his work on plants and air, *Vegetable Staticks*

1733 John Kay patents the so-called flying shuttle, a breakthrough in weaving, in England

1733–37 Charles Dufay begins publishing six memoirs on electricity

1735 Carolus Linnaeus's *Systema naturae* ("System of Nature")—his system of classification of organisms—is published

1736 Pierre de Maupertuis leads expedition to Lapland to measure the length of a degree; proves Earth is flattened at poles

1737 Calcutta earthquake kills 300,000

1738 Daniel Bernoulli's *Hydrodynamica* is published, explaining relationship between pressure and velocity of fluids; Bernoulli effect

1738 Maupertuis publishes results of his expedition, which support Newton's theory about the shape of the Earth

1738 Voltaire's *Eléments de la philosophie de Newton* is published

1739 Royal Society of Edinburgh founded (receiving a charter in 1783); scientific society founded also in Stockholm, later called Kungliga Svenska Vetenskapsakademien

1739–40 David Hume, Scottish philosopher, publishes *A Treatise of Human Nature,* in which he tries to apply experimental method to human psychology

1742 Royal Danish Academy of Sciences and Letters is founded

1742 Anders Celsius invents the Celsius temperature scale

1743 Benjamin Franklin helps establish the American Philosophical Society at Philadelphia

1743 Jean d'Alembert publishes *Traité de dynamique* ("Treatise on Dynamics"), expanding on Newton's laws of motion

1744 Maupertuis formulates principle of "least action" in physics

1745 Buffon proposes that the Earth formed when a comet collided with the Sun

1746 Denis Diderot publishes *Pensées philosophiques,* postulating that the presence of order in nature proves the existence of God

1746 Leonhard Euler works out the mathematics of the refraction of light

1746 Leiden Jar invented, facilitating the study of electricity

1746 William Watson publishes *Experiments on the nature of electricity*

1747 Jean d'Alembert presents first general use of partial differential equations in mathematical physics

1748 Mikhail Lomonosov formulates the laws of conservation of mass and energy

1748 Julien de La Mettrie publishes *L'homme machine*, describing humans as machines without free will

1748 John Needham and Buffon think they prove spontaneous generation

1749 Buffon formulates modern definition of species and publishes his *Théorie de la terre* ("Theory of the Earth") in the first volume of his *Histoire naturelle...*, a 55-year, 44-volume project that popularizes the study of nature

1749 Denis Diderot publishes *Lettre sur les aveugles*, setting forth ideas on materialism and on human dependence on senses and his theory of variability and adaptation

1749 Emilie du Châtelet completes translation of Newton's *Principia* into French from Latin. It is published in 1759 with a preface by Voltaire and becomes widely influential

1750 Thomas Wright publishes *An Original Theory and New Hypothesis of the Universe*

1750–60 Approximate beginning of the Industrial Revolution in England

1751 First volume of the *Encyclopédie* is published; last volume published 1772, total 17 volumes of articles and 11 of illustrations; 5 supplementary volumes follow in 1776–77

1751 Benjamin Franklin publishes *Experiments and Observations on Electricity*

1751 Carolus Linnaeus publishes *Philosophia botanica* ("Botanical Philosophy")

1751 Pierre de Maupertuis publishes *Système de la nature* ("System of Nature"), theoretical speculation on heredity and the origin of species by chance

1752 Benjamin Franklin's kite experiment

1752 Jean Guettard publishes short booklet explaining his ideas about geology and rock strata

1753 Carolus Linnaeus publishes *Species plantarum* ("Species of Plants")

1753 British Museum chartered

1753 Immanuel Kant formulates ideas on evolution of the universe from primeval nebulas

1754 Charles Bonnet publishes *Recherches sur l'usage des feuilles des plantes*, on nutritional value of plants; Etienne Bonnet publishes *Traité de sensations*, claiming that knowledge only reaches us through the senses

1755 Kant formulates dust cloud theory of origin of Solar System and proposes that nebulas are star systems like the Milky Way in

	his *Allgemeine Naturgeschichte und Theorie des Himmels* (*General Natural History and Theory of the Heavens*)
1755	Earthquake in northern Persia kills 40,000 and Lisbon earthquake kills 60,000
1756	Joseph Black publishes *Experiments upon magnesia . . .*, first detailed account of chemical action and the first work of quantitative chemistry
1756	Voltaire's "Essay on the Customs and Manners of Nations" published
1759	Caspar Friedrich Wolff publishes *Theoria generationes* ("Theory of Generation"), refuting homunculus theory
1760	Giovanni Morgagni establishes pathology as a science
1760s	James Cook surveys coasts of Labrador and Newfoundland; first of scientific navigators; three scientific voyages to Pacific: 1768–71, 1772–75 and 1776–79
1761	Joseph Black discovers latent heat
1761	First observation of transit of Venus organized
1761	Morgagni does first important work in pathological anatomy (on the causes of diseases)
1762	James Bradley compiles new star catalog of 60,000 stars
1762	Catherine II the Great becomes ruler of Russia
1764	Voltaire publishes *Dictionnaire philosophique*, argues against Great Chain of Being
1764	Charles Bonnet develops preformation theory
c. 1766	Lunar Society founded in England. It is a private group that draws members from both science and industry
1766	Solar eclipse observed by James Cook during his survey of the coast of Newfoundland
1766	Cavendish announces his work on hydrogen to the Royal Society (discovery of hydrogen)
1766	Joseph Priestley meets Benjamin Franklin in London
1767	Priestley publishes his *History of Electricity*
1768–71	First of James Cook's voyages to the Pacific
1769	James Watt invents the first practical steam engine
1769	Worldwide observations of the second transit of Venus of the century
1770	In America, British soldiers fire on colonists in the Boston Massacre. The population in the American colonies is now nearly 2.2 million
1772–75	Cook's second voyage to the Pacific
1773	Karl Wilhelm Scheele discovers oxygen a year before Priestley does but the work is not published until 1777

1773 Boston Tea Party: Angry American colonists dump British tea into Boston waters in an act of rebellion
1774 Priestley discovers oxygen (actually Scheele isolated it first but did not publish)
1775 Beginning of the American Revolution
1776–79 Cook's third and last voyage to the Pacific
1776 American Declaration of Independence signed; Revolution continues. The largest city in America, Philadelphia, now has a population of nearly 40,000
1777 Scheele publishes *A Chemical Treatise on Air and Fire* (describing his discovery of oxygen)
1779 Lavoisier names oxygen
1781 Charles Messier compiles a catalogue of more than 100 nebulas
1781 James Watt, who has learned a great deal about heat from his friend the chemist Joseph Black, perfects his steam engine. It is another major step in the Industrial Revolution
1781 William Herschel discovers Uranus
1782 John Goodricke announces his observations of a "variable" star, Algol
1782 Oliver Evans works on an early version of a locomotive
1783 The American Revolution ends and Great Britain recognizes the independence of the United States
1783–84 Henry Cavendish demonstrates that when hydrogen burns it produces water, proving that water is a combination of two gases, not itself an element as thought by the Greeks
1784–89 Charles Augustin de Coulomb writes his memoirs on electricity and magnetism
1784 James Watt patents a locomotive two years after Oliver Evans had patented a similar device
1785 Cavendish finds evidence of the existence of argon (confirmed in following century by Ramsay)
1786 Lavoisier names hydrogen (discovered by Cavendish 20 years earlier)
1787 Abraham Werner, primarily a teacher of great influence, publishes a 28-page booklet outlining his ideas
1788 James Hutton first publishes his theory of Earth in the Royal Society of Edinburgh's *Transactions*
1789 Antoine Lavoisier publishes *Traité élémentaire de chimie*
1789 Enraged French citizens attack the Parisian prison known as the Bastille, marking the beginning of the French Revolution
1791 Luigi Galvani publishes *De viribus electricitatis in motu musculari commentarius*
1793 In America, the inventor Eli Whitney invents the cotton gin

1793 Louis XVI is executed: Reign of Terror begins in France, lasting for two violent and bloody years

1794 Lavoisier executed

1794 In England rioters afraid that the violence of the French Revolution will strike their country attack supporters of the French revolutionaries, including Joseph Priestley, an outspoken supporter of the Revolution, and he flees to the United States

1795 A new French constitution establishes the Directory

1795 Hutton publishes his theory with new materials in two volumes, *Theory of the Earth*

1795 Georges Cuvier appointed assistant to the professor of comparative anatomy at the Museum of Natural History in Paris

1796 Cuvier begins his famous lectures at the Museum of Natural History in Paris

1796 Pierre Simon Laplace proposes the nebular hypothesis

1796 English physician Edward Jenner pioneers the use of vaccination against smallpox

1798 Benjamin Thompson (Count Rumford) publishes his *Experimental Inquiry Concerning the Source of Heat Excited by Friction*

1798 Jenner publishes *An Inquiry into the Causes and Effects of the Variolae Vaccinae . . .*

1798 Eli Whitney builds a factory in New Haven, Connecticut to mass-produce firearms

1799 Alessandro Volta sends his famous letter to the President of the Royal Society, Joseph Banks, titled, "On the electricity excited by the mere contact of conducting substances of different kinds"

1799 Napoléon Bonaparte seizes power in France

1800 Cuvier publishes his *Leçons d'anatomie comparée* ("Lessons on Comparative Anatomy")

1800 Herschel builds 40-foot-long reflecting telescope

GLOSSARY

astronomical unit The distance from the Earth to the Sun, 93 million miles, often used as a unit of measurement in astronomy (abbreviation: AU)

atom Leucippus (Greece, 8th century B.C.) said atoms were the smallest indivisible unit of matter, differing only in shape and size. Today's ideas about atoms originate from models developed by Ernest Rutherford and Niels Bohr

caloric Term used in the 18th century to describe heat seen as a fluid, used by Lavoisier and others

calx Oxide (18th-century term)

catastrophist geology The school of thought that held that the Earth's geological features were the product of sudden, catastrophic events, such as floods, earthquakes and volcanoes (compare *uniformitarian geology*)

chemistry Today, the scientific study of the composition, structure, properties and reactions of matter, especially at the level of atomic and molecular systems. In the 18th century, the study of mixtures of substances and the results of changing the temperature of substances (compare *physics*)

dephlogisticated air Oxygen (term used by Joseph Priestley in the 18th century)

element Defined in 1787 by Antoine Lavoisier as a substance that is chemically "simple"; in other words, one that cannot be further decomposed

epigenesis Series of distinct phases through which an embryo passes

fixed air Eighteenth-century term for carbon dioxide

galaxy A large group of stars, cosmic dust and gas bound together by gravitation into an astronomical system

gravity The natural force that makes objects having mass attract each other

nebula (plural: nebulas) A large cloud of interstellar gas and dust. The term originally was used for any object in the heavens that appeared fuzzy through the telescope

nebular hypothesis An early theory of the origin of the Solar System (Kant, Laplace) that proposed that it was formed from a cloud of material collapsing under the effects of gravitation

Neptunism Theory of geology that holds that all rocks are laid down by water and sediment (compare *Plutonism*)

parallax An apparent shift in position of an object viewed from two different locations, for example, from the Earth at two different positions on opposite sides of its orbit (achieved by observing the objects at six-month intervals). The greater the apparent change in position, or annual parallax, the nearer the star

periodical Occurring at regular intervals, having periods or repeated cycles

phlogiston According to 18th-century belief, one of three varieties of "earth," a combustible material lost when a substance burns

physics The scientific study of matter and energy and interactions between them; originally, the study of all natural things (compare *chemistry*)

Plutonism The theory, common in 1800, that the origin and nature of all rocks (and thus, the crust of the Earth) is explainable by the effects of heat (compare *Neptunism*)

teleology Use of predetermined purpose to explain the universe; the idea that future events control the present, that is, that "A is so in order that B might be so," that nature or natural processes are directed toward an end or shaped by a purpose

triangulation A system used to measure distance through the use of known facts about the relationships of angles and distances in triangles

uniformitarian geology The school of thought that held that geological features are the product of natural forces operating steadily and slowly over long periods of time (compare *catastrophist geology*)

FURTHER READING

ABOUT SCIENCE:

Cole, K. C. *Sympathetic Vibrations: Reflections on Physics as a Way of Life.* New York: William Morrow, 1985. Well-written, lively and completely intriguing look at physics presented in a thoughtful and insightful way by a writer who cares for her subject. The emphasis here is primarily modern physics, concentrating more on the ideas than the history.

Ferris, Timothy, ed. *The World Treasury of Physics, Astronomy and Mathematics.* New York: Little, Brown, 1991. Anthology of mostly modern physics but includes some general papers on the philosophy of science as well as some delightful poetry on physics. Sometimes difficult but well worth browsing through.

Fisher, David E. *The Creation of the Universe.* Indianapolis: Bobbs-Merrill, 1977. Easy-to-follow, enthusiastic narrative aimed at the younger reader.

Fleisher, Paul. *Secrets of the Universe: Discovering the Universal Laws.* New York: Atheneum, 1987. Well-written and engrossing look at physics aimed at the younger reader but enjoyable and informative for anyone interested in science. Includes excellent discussions of the work of Galileo, Newton and others, as well as general laws of physics.

Gonick, Larry, and Art Huffman. *The Cartoon Guide to Physics.* New York: Harper Perennial, 1991. Fun, but the whiz-bang approach sometimes zips by important points a little too fast.

Hann, Judith. *How Science Works.* Pleasantville, N.Y.: Reader's Digest, 1991. Lively, well-illustrated look at physics for young readers. Good, brief explanations of basic laws and short historical overviews accompany many easy experiments readers can perform.

Hazen, Robert M., and James Trefil. *Science Matters: Achieving Scientific Literacy.* New York: Doubleday, 1991. A clear and readable overview of basic principles of science and how they apply to science in today's world.

Holzinger, Philip R. *House of Science*. New York: John Wiley & Sons, 1990. Lively question-and-answer discussion of science for young adults. Includes activities and experiments.

Morrison, Philip, and Phylis Morrison. *The Ring of Truth: An Inquiry into How We Know What We Know*. New York: Random House, 1987. Companion to the PBS television series. Wonderful and highly personal explanations of science and the scientific process. Easy to read and engrossing.

Rensberger, Boyce. *How the World Works: A Guide to Science's Greatest Discoveries*. New York: William Morrow, 1986. Lucid and readable explanations of major scientific concepts, terms explained in A-to-Z format and excellent discussions of the scientific method and how science works.

Trefil, James. *1001 Things Everyone Should Know about Science*. New York: Doubleday, 1992. Just what the title says—and well done.

Walker, Jearl. *The Flying Circus of Physics with Answers*. New York: John Wiley & Sons, 1977. A by-now classic collection of problems and questions about physics in the everyday world. Little history but much fun.

ABOUT THE HISTORY OF SCIENCE:

Asimov, Isaac. *Asimov's Chronology of Science and Discovery*. New York: Harper and Row, 1989. Lively chronological view of science, year by year. Written with Asimov's usual verve. Good for fact-checking and browsing.

Boorstin, Daniel J. *The Discoverers*. New York: Random House, 1983. Its size may be intimidating—over 700 pages—but this is a wonderfully lively, thoughtful and absorbing look at the history of humankind's search to know itself and nature. Aimed at the general reader.

Brooke, John Hedley. *Science and Religion: Some Historical Perspectives*. Cambridge: Cambridge University Press, 1991. This higher-level book is sometimes difficult, but it gives a well-presented and insightful look at the relationship between science and religion throughout history.

Gillespie, Charles C. *The Edge of Objectivity*. Princeton, N.J.: Princeton University Press, 1960. Intriguing higher-level look at science, its history and philosophy.

Hays, H. R. *Birds, Beasts, and Men: A Humanist History of Zoology*. New York: G. P. Putnam's Sons, 1972. Readable and well-organized narrative, but old and may be hard to find.

Hellemans, Alexander, and Bryan Bunch. *The Timetables of Science.* New York: Simon and Schuster, 1988. An easy-to-read, year-by-year chronology of science. Good for fact-finding or just browsing. And the "overviews" are nicely done, adding historical context.

Magner, Lois N. *A History of the Life Sciences.* New York: Marcell Dekker, 1979. A good, readable overview though marred somewhat by awkward organization.

Mason, Stephen F. *A History of the Sciences,* revised edition. New York: Collier Books, 1962. Originally published in 1956, this book is older than Ronan's but is a solid standard history of science.

Motz, Lloyd, and Jefferson Hane Weaver. *The Story of Physics.* New York: Avon Books, 1989. Occasionally rough going but well worth a try for a good, strong narrative history of the theories, ideas, experiments and people of physics.

Ronan, Colin. *The Atlas of Scientific Discovery.* New York: Crescent Books, 1983. A slim coffeetable-type book but well-written and thought out overview with illustrations.

Ronan, Colin A. *Science: Its History and Development Among the World's Cultures.* New York: Facts On File, 1982. A good, readable comprehensive overview of the history of science from the ancients to the present.

Sambursky, S. *The Physical World of the Greeks.* Princeton, N.J.: Princeton University Press, 1956. Reprinted in paperback, 1987. For the reader wishing to delve much more in detail into the classic physics of the great Greek philosophers. Tough going in spots for the younger or general reader.

Thiel, Rudolf. *And There Was Light: The Discovery of the Universe.* New York: Alfred A. Knopf, 1957. A little difficult to find but an interesting narrative history of astronomy.

SCIENCE IN THE EIGHTEENTH CENTURY:

Adams, A. B. *Eternal Quest.* New York: G. P. Putnam's Sons, 1969. Readable, conversationally told profiles of scientists such as Linnaeus, Buffon and others, including biographical background and historical context.

Asimov, Isaac. *Eyes on the Universe.* Boston: Houghton Mifflin Company, 1975. A good starting place for readers wanting to know more about the early history and development of telescopic instruments.

———. *The Search for the Elements.* New York: Basic Books, 1962. Concisely told tale of the search to define what an element is, to discover the elements in nature and to organize them logically.

Bowler, Peter J. *Evolution: The History of an Idea*, revised edition. Berkeley, Calif.: University of California Press, 1989. An excellent look at the history of evolutionary theory. Includes many of the subjects included in this book. Highly thoughtful and informative.

Bronowski, J., and Bruce Mazlish. *Western Intellectual Tradition.* New York: Harper and Row Publishers, 1960. A fascinating and thoughtful overview of the growth of modern thought in science and philosophy.

Butterfield, Herbert. *The Origins of Modern Science.* New York: Macmillan Company, 1958. Includes, along with much other valuable information, a good discussion of the problems in chemistry that account for its slow development as a science.

Ferris, Timothy. *Coming of Age in the Milky Way.* New York: William Morrow and Company, 1988. Easy-to-read and enjoyable history of astronomy in practice and theory.

Gould, Stephen Jay. *The Flamingo's Smile.* New York: W. W. Norton and Company, 1985. The always readable Gould here includes chapters on the Great Chain of Being and the ideas and work of Maupertuis, along with other assorted essays on natural history and evolutionary theory.

———. *Hen's Teeth and Horse's Toes.* New York: W. W. Norton and Co., 1983. Has a good chapter on Hutton and his ideas. Also chapters on Cuvier and early geography.

———. *Ever Since Darwin: Reflections in Natural History.* New York: W. W. Norton and Company, 1977. Includes an enlightening discussion on early ideas about preformation.

Hays, H. R. *Birds, Beasts and Men.* New York: G. P. Putnam's Sons, 1972. Readable and well-organized narrative subtitled "A Humanist History of Zoology," but it is older and may be hard to find.

Mayr, Ernst. *The Growth of Biological Thought: Diversity, Evolution, and Inheritance.* Cambridge, Mass.: Harvard University Press, 1982. Impressive, if sometimes heavy-going, look at the history of evolutionary theory by one of the world's leading experts.

Milner, Richard. *The Encyclopedia of Evolution: Humanity's Search for Its Origins.* New York: Facts On File, 1990. Engrossing for hunting facts or browsing, one of the few books of its kind that is also completely rewarding, entertaining and informative as just plain reading.

Partington, J. R. *A Short History of Chemistry.* New York: Dover Publications, Inc., 1989. Originally published in 1937 and revised in 1957. Still one of the better brief histories.

Richardson, Robert S. *The Star Lovers.* New York: Macmillan Company, 1967. Although a little technical at times, well worth a try for its nice integration of science and narrative as it looks at the lives and thoughts of many of the world's greatest astronomers.

Roberts, Gail. *The Atlas of Discovery.* New York: Gallery Books, 1989. With an introduction by Sir Francis Chichester. Colorful detailed maps outline the adventures of the world's greatest explorations on land and sea. A lively text complements the maps in this well-produced book.

von Baeyer, Hans. *Rainbows, Snowflakes and Quarks: Physics and the World Around Us.* New York: McGraw-Hill Book Company, 1984. An easy-to-read and informal collection of essays. Has a very good short look at Rumford and his work.

ABOUT SCIENTISTS:

Abbott, David, ed. *The Biographical Dictionary of Scientists: Astronomers.* New York: Peter Bedrick Books, 1984. Short entries from A to Z, including an extensive glossary and line diagrams. Dry and a little difficult but a good resource.

———. *The Biographical Dictionary of Scientists: Physicists.* New York: Peter Bedrick Books, 1984. Like Abbott's dictionary of astronomers, a reliable reference, though somewhat tough going.

Asimov, Isaac. *Asimov's Biographical Encyclopedia of Science and Technology*, second revised edition. Garden City, N.Y.: Doubleday and Company, 1982. The somewhat unusual chronological and nonalphabetical entry system takes some getting used to, but overall a lively and typically opinionated Asimov approach makes for fascinating reading as well as basic fact gathering.

Crowther, J. G. *Scientists of the Industrial Revolution.* London: The Cresser Press, 1962. Highly readable account of the lives and work of Joseph Black, James Watt, Joseph Priestley and Henry Cavendish and some of their contemporaries.

Meadows, Jack. *The Great Scientists.* Oxford: Oxford University Press, 1987. Easy to read, well presented and nicely illustrated.

INDEX

Boldface numbers indicate major topics. *Italic* numbers indicate illustrations.

E

F

G

O

P

Q

R

S